Classification, Disease and Evidence

History, Philosophy and Theory of the Life Sciences

Volume 7

For further volumes:
http://www.springer.com/series/8916

Philippe Huneman • Gérard Lambert
Marc Silberstein

Editors

Classification, Disease and Evidence

New Essays in the Philosophy of Medicine

 Springer

Editors
Philippe Huneman
IHPST
CNRS/Université Paris I/ENS
Paris, France

Gérard Lambert
Centre Cavaillès
Ecole Normale Supérieure
Paris, France

Marc Silberstein
Éditions Matériologiques
Paris, France

ISSN 2211-1948   ISSN 2211-1956 (electronic)
ISBN 978-94-017-8886-1   ISBN 978-94-017-8887-8 (eBook)
DOI 10.1007/978-94-017-8887-8
Springer Dordrecht Heidelberg New York London

Library of Congress Control Number: 2014948778

Printed on acid-free paper

Springer is part of Springer Science+Business Media (www.springer.com)

Contents

Introduction: Surveying the Revival in the Philosophy of Medicine

The Emergence of the Philosophy of Medicine as an Academic Discipline

The Philosophy of medicine is currently experiencing a fascinating revival. Journals are publishing more and more papers about the field, and textbooks for the discipline have begun to appear in the last decade (Sadeg-Zadeh 2012; Gifford 2011). Granted, medicine has always been an object of concern for philosophers, either (to take examples from antique tradition) under the general heading of "embodiment" – the relationship between body and mind, the relative weights of vitalism and mechanism, etc. – or in the mode of a generalized use of the terms "health" and "disease" as a scheme for elaborating normative judgments. Nietzsche's use of the terms health and disease to condemn or praise forms of life and civilization, as well as Marx's or Freud's diagnosis of diseases in modern civilization, instantiated this form of a constitutive relationship between philosophy and medicine, which is not at all a concern for what we currently call the philosophy of medicine.

More generally, it is easy to discern important episodes in the history of this proximity or kinship between the two disciplines: Descartes thought that medicine, as one of the highest branches of the tree of knowledge, could be a terminus ad quem of scientific investigation; and long before him, Socrates and other Greek philosophers sometimes viewed themselves as physicians of the soul. Recently, Wittgenstein and his epigones conceived of philosophy as a sort of therapy of meaning; and here again, the health/disease conceptual pair seems to play an even more radical or foundational role than the traditionally philosophical opposition between the true and the false.

But the recent philosophy of medicine can be seen as a specific and autonomous subfield of philosophy – which could perhaps be specified in a difference from the "medical philosophy" represented by the aforementioned tradition, exactly in the way one classically opposes "philosophy of biology" to "biological philosophy." As such, the philosophy of medicine is structured by a set of recent questions whose importance and nature stem from both philosophy and medicine.

1. First, since the nineteenth century there has been a continued questioning of the meaning of health and the nature of disease. This was a debate within medicine itself, and the birth of clinical medicine, which has been so important for the paradigm of modern medicine, was accompanied by a debate on the very nature of disease and the relationship between physiological and pathological domains. Famously, Claude Bernard (1859) borrowed from Broussais the idea that the very difference between health and disease is quantitative – meaning that a pathological state is a variant of a physiological state. In one of the milestones of what could be called the "non-analytic philosophy of medicine", *The Normal and The Pathological* (1959), the philosopher Georges Canguilhem discusses this position, arguing that normativity is a property of the living body as such (which implies that there is a qualitative difference between normal and abnormal) and that norms will always be seen somehow in context – as living activity is always likely to define and change its own norms. The nature of health and disease is therefore a longstanding issue for philosophers, and of course satisfies the philosopher's taste for foundational issues: whereas medicine investigates diseases, philosophy examines what "to be a disease" means. Some of the most important contributions to this question (e.g. Boorse's theory of health (Boorse 1975) or Jerome Wakefield's idea of disease as harmful dysfunction (Wakefield 1992)) subscribed to such idea of a division of labor between the philosopher and the physician. Today, general thought on health and disease in the past century has been deeply shaped by three independent sources:

1. The idea of conceptual analysis (and thus the inclination towards capturing concepts in terms of necessary and sufficient conditions) – which came from the tradition of analytical philosophy, and which shaped the methods of Boorse, Wakefield and many others. Such a philosophical project was stimulated by two things:

    (a) The radical critique of medicine and psychiatry which arose in the 1960s, and whose general target has been the social and cultural dimension of any value judgement or norm – a critique that can be found in many theoretical perspective, be it the "archaeology of knowledge" developed by Foucault, the antipsychiatry led by Cooper and Laing, or the radical critique of modern medicine developed by Ivan Illich and his followers in the 1970s.

    This latter debate directed the question of the definition of disease towards the specific debate over the possibility of a purely biological, value-free, definition of health.

    (b) The change in nature of the predominant diseases in Western countries, shifting from acute infectious diseases to chronic diseases (e.g. cardiovascular diseases, diabetes, cancer, etc.). Responding to this new situation, medicine became more and more preventive, treating diseases before the onset of symptoms, and blurring the boundary between disease and risk factors (e.g. is hypertension a disease?). In the Middle Ages, a diagnosis of leprosy was reason enough to banish the leper from society. In modern welfare states, a diagnosis may also result in one being locked up; but health is now deemed as a right, and illness can provide social advantages. The passionate reactions elicited by the editors of the *British Medical Journal* asking their readers to vote for the "top-10 non-diseases" (Smith 2002) are linked to the social implications of disease and show that this question is not simply a philosophical one – even though philosophers are needed to correctly formulate the question.

2. Second, the advances in biology that, in a word, tied both the causes of diseases and the cure of diseases to the lower level of cellular, genetic and molecular processes raised philosophical issues concerning both the nature of medical knowledge and the ontological nature of its object. The emergence of bacteriology, biochemistry and parasitology in the beginning of the twentieth century was a first step in this process – with of course the discovery of the role of microbes in infectious diseases; then came genetics, the investigation of the patterns of heritable diseases, immunology, and finally molecular biology. After the discovery of DNA and all that followed, it became possible to both identify some diseases by a mutation in a DNA sequence (e.g. beta-thalassemia or sickle-cell anaemia, the first "molecular disease" as Linus Pauling termed it (Pauling et al. 1949)), and of course to design specific tests – more generally, to trace pathological conditions back to abnormal mechanisms (which of course leaves the question open about what "abnormal" means). This raised new questions – especially concerning the status of medical activity in relation to biology, pharmaceutics, and hospitals.

3. Third, the salience of statistical schemes and methods in assessing etiologies and therapeutics. Much has been written about the statistical or "probabilistic revolution" (e.g. Krüger et al. 1987; Hacking 1975), which started in the eighteenth century and wholly transformed medicine in the twentieth century while also opening up new areas of philosophical investigation. For instance, the rise of epidemiology – due both to better access to population data, and to better statistical tools which make it possible to track the origin and diffusion patterns of diseases, and therefore to establish correlations – has created methodological and then philosophical questions concerning statistical inferences. A canonical example of causal Judgement in medicine, "smoking causes lung cancer," is indeed based on a statistical consideration of a set of cases. Such judgments clearly raise basic problems with the relationship between causation and statistical correlation. (It is interesting to remember that one of the first studies to show a causal link between smoking and lung cancer (Doll and Hill 1954; Doll et al. 2004) was objected to by the founding father of modern statistics, Ronald Fisher (himself a heavy smoker), who speculated that it spuriously detected a causal link when there was only a common cause (e.g. common genetic predisposition) between two facts). More generally, given that there is variation between individuals, and as there is a multiplicity of factors involved in any pathological event, inferring judgments about the general validity of a cure or even the causes of a disease is hardly possible on the basis of a single case history, and therefore requires the collection and comparison of many cases. The usual methodology for clinical trials therefore always rests on statistical methods – even though there are hundreds of them, and their use and legitimacy raise methodological and epistemological issues that are indeed part of the philosophy of medicine. An important area of the philosophy of medicine is therefore oriented towards asking questions about evidence (for either causal ascription or therapeutics efficiency) within a statistical framework.

Numerous investigations into the methods of identifying and explaining diseases, as well as assessing cures, have been conducted in the last decade – some of them very formal (Lucas 2001; Nikovski 2000), using tools like Bayesian networks (Spirtes et al. 2000); others including a more descriptive component that sometimes touches upon the sociology of medicine and its history (e.g. Barrett 1995). Many of these studies have focused on randomized control trial (RTC), a very generalized tool for testing drugs in their many varieties. They have been stimulated by the emergence of so-called evidence-based medicine (Howick 2011; Guyatt et al. 1996), whose ambitions are to make medicine both more efficient and more rigorous, and which massively uses RTCs. Thus, a crucial question is to make sense of the kind of knowledge brought about by statistically established correlations – especially when no other data can be used to control them. As a result, some philosophers argue that no purely statistical knowledge can provide us with a full causal explanation of either a disease or the efficiency of a drug if it is not backed up by experiment-based evidence of a potential mechanism that underpins the putative causal relations between a disease and an agent, or a drug and relief (Russo and Williamson 2007). However, other philosophers tend to defend the validity of RTCs and other statistical methods in providing causal knowledge of medical facts. Yet, since a physician cures an individual and not a population or an average person (as Aristotle famously noticed a long time ago), another question is how one can apply cohort statistics to a single person?

For all these reasons, the philosophy of medicine ended up including many more questions than the foundational problem of the nature of diseases. Research questioning the relationship between medicine and biology – as well as the role of distinct biological theories (molecular, genetic, evolutionary) in our understanding of disease – emerged within the field of the philosophy of science.

On the other hand, in the overall field of the philosophy of medicine, many studies which focus on the ambiguous status of psychiatry – which is a recent branch of medicine, since it emerged in the nineteenth century (Goldstein 1987; Scull 1975) and still struggles with theorists who challenge the idea that insanity is a genuine disease – have been interested in both the controversial notion of "mental illness" and the divide between the conflicting theories intended to address it (psychodynamic, systemic, genetic, molecular, cognitivist, neuroimaging, etc.). This is a very attractive topic for a philosopher, given that any step towards the formulation of a theory in psychiatry may of course raise deep philosophical questions (concerning normativity, values vs. nature, mental states, cognition, mind/body, free will, causality, reductionism, etc.). The DSM handbook – which was published in the 1950s and was a research tool rather than a clinical one in its early conception – became the most used and widespread book for help in the diagnosis of patients suffering from putative mental conditions. This, along with its ambition to be almost theory-neutral, has therefore attracted a lot of attention. Indeed, the public debates surrounding the preparation of DSM 5 (published in May 2013; Demazeux and Singy 2014) involved many of these deep philosophical issues, and were discussed by psychiatrists, biologists, and philosophers. And before that, the DSM 3 was largely influenced by philosophical views on health and disease – including Wakefield's harmful dysfunction concept.

Questions in the Philosophy of Medicine

Even though this is a burgeoning and growing field, the philosophy of medicine seems to be oriented towards four very general (interdependent) questions that somehow mirror the three conditions we sketched: (a) What is a disease and what is health? (b) How do we (causally) explain diseases? (c) How do we assess and choose cures/therapy? (d) And how do we distinguish diseases, i.e. define classes of diseases and recognize that an instance X of disease belongs to a given class of A?

Causation and Causal Explanation

As such, the question of explanation (b) encompasses a huge set of problems: identifying causal factors, weighing them, specifying what is distinctly medical in an explanation, disentangling causation from spurious causation (correlation) in statistical data, etc. Many of these problems are very general problems of explanation within the sciences. But at least two things are proper to medicine here. First, the answers one gives are not independent from the position one takes regarding question (a) – that is, if a disease is a deviation from normal functioning, it will require a different kind of explanation than if it were a state of being heterogeneous to healthy physiology. Second, as Aristotle emphasized a long time ago, the physician cures an individual, not a type (of disease) or a class (of diseased individuals) – and therefore explanations, causation, and all related notions should be defined from this perspective. Especially since humans are biological, sociological, and psychological entities at once, all medical cases occur at the intersection of many regimes of causality (sociological, psychological, biological, chemical, etc.), and span many levels and scales (bacteriology, cells, etc.); thus the weight of the relative impact of causal processes concerning a particular individual disease requires careful methodological examination. Many questions regarding the status of medicine – its relation to biology, physiology, chemistry – and, within medicine, the relationships between clinical medicine, hospital care, and laboratory activity are concerned with this specific feature of medical activity at the crossroads of heterogeneous logics and heterogeneous causal processes.

In this regard, let us take the opportunity of this Introduction to put issues concerning medical explanation in the light of the history of medicine. A traditional medical perspective on causal explanation is the idea that diseases have, in principle, two heterogeneous types of causes: some of them are "predisposing" causes, meaning that they are traits proper to an individual or to her way of life that enter into the explanation of the disease but do not necessarily and directly cause it themselves; others are "triggering" causes, meaning that they cause the disease by acting on these specific "predisposing" causes. In his *Médecine Clinique* (1801) – which accompanied his work on nosology (*Nosographie philosophique*, 1800), and was a major treatise for all the leaders of the inchoative "clinical medicine" in early nineteenth century Paris – Philippe Pinel says that the physician must "look for the predisposing and exciting causes" of a disease in all cases. These should be looked for: "(1) Within the job and the way of life of the patient; (2) In the accidents prior to the current illness, in the previous state of health; (3) Sometimes, among the diseases which relatives of the patient have suffered" (Pinel 1813, 5). This is a well established medical explanatory scheme: in the eighteenth century, François Boissier de Sauvages stated that the "efficient causes," which "effectively produce the illness," differ from the "conditions without which it couldn't occur" (Sauvages (1772), I, §155, p.187). In the same period, Cullen asserted that although hypochondria is the result of moral causes, the "bodily temperament determines those causes to produce their effect sooner or later" (Cullen (1784), §1229). Regarding epilepsy, he juxtaposed the "collapsing causes" – hard bleeding (§1301), terror (§1302) – with the "predisposing causes" – such as "motility," which displays the state of mind (§1307) and consists in "a more or less high degree of sensibility or irritability" (§1311). Also in this period, Whytt (1765) identified two kinds of causes: "predisposing causes" (ch.III), which are divided into the weakness of a particular organ and the excessive delicacy of the whole nervous system (§XXXI); and "occasional causes" (ch.IV), which can be either local or general (i.e. in blood (§LIII)). This divide was not always very determined: at the end of the century, Crichton (1798) included passions within the "exciting" causes, but nothing precluded them from being "predisposing" causes if they could act in a long-lasting way. The classical notions of "temperament" (or, earlier on, "constitution"), especially within the solidist medicine of previous centuries, were also a way of describing the sets of "predisposing" causes. To some extent, advances and changes in medical explanations can be seen as providing new ways of understanding these two regimes of causation and their expression: nowadays, genes are likely to define the "predisposing" causes, and microbes, infections, or life events (stress, for example) define the "exciting" causes; yet, in some cases, recurring conditions of life (e.g. reccuring high degrees of stress, histories of early child abuse, etc.) provide the predisposing causes. This is somewhat reminiscent of Ernst Mayr's distinction between proximate and ultimate causes in biology (Mayr 1961) – a dual system picked up by so-called Darwinian medicine (Williams and Nesse 1991; Methot 2011) where proximate causes are the exciting causes sensu traditional physicians and ultimate causes are evolutionary and almost entirely represented by genes, which explains why humans as a species are susceptible to certain diseases and not to others (Nesse 2001). The whole project of evolutionary medicine can of course be seen as a systematic development of such an evolutionary take on the traditional dual system of causes inherited from the medical tradition.

  This illustrates the fact that many contemporary philosophical views about explanations in medicine could be used to make sense of this very general explanatory divide that physicians have traditionally used to understand their own practice of looking for etiologies. Recent notions, often of a probabilistic nature, such as "risk" ("risk group", "predictor", etc.), have a modern, post-"probabilistic-revolution" approach to this old idea of "predisposing causes." Interestingly, medical disciplines such as epidemiology or medical genetics can be understood within the general structure of modern medicine, by specifying their contribution to the definition of each regime of causation in the double etiology scheme.

Nosology and Ontology

Given that the first objective of a physician is to establish a diagnosis – namely, to determine under which concept of disease the case under consideration falls – medicine necessarily needs a classification of diseases. Classifications as such raise certain issues of principle: are classes objective concepts, or just ways of grouping various facts in a fashion that serves our practical (here: therapeutic) purposes? What are the crucial properties or facts that one should use in classification? Are classifications of various natures dependent on these criteria, and should they be used to assess other classifications (for example, are the recent medical classifications based on networks of disease genes (Barabasi et al. 2011) supposed to match up with traditional classifications?). Moreover, if one single criterion is unable to classify all diseases, there is no possible unity in any type of classification of diseases.

Classification in general is laden with metaphysical and philosophical issues. Indeed, in eighteenth century biology, naturalists such as Linnaeus, John Ray, or Buffon disputed about both the importance of the traits they would use to distinguish and assemble individuals, and the realism of the ensuing classificatory systems – namely, do they "carve nature at its joints", and should they even try to do so? Positions then ranged from extreme nominalism – notably Buffon's, who held that only individuals exist in nature, and that species, genera, or families are just names which are useful for us in grouping individuals according to our explanatory and pragmatic goals – to Linnaeus, who thought that even though the choice of his classification criteria had no consideration in regard to the biological importance of the organs themselves, his table of species would still match the repartition of species as they were originally created. Granted, classifying diseases shares in the problems of classification in general, and in this respect, it is interesting to notice that Sauvages, author of an important nosography (Sauvages 1772), also wrote a work about the classification of plants – the *Methodus foliorum, seu plantae florae monspeliensis, juxta foliorum ordinem, ad juvandam specierum cognitinem, digestae.* (A Method to know plants by their leaves). From Thomas Sydenham to Philippe Pinel, diseases were seen as natural entities difficult to recognize because they were in a sense 'corrupted' by their instantiation within a patient, and nosology was a quest for a natural and hierarchical order of fixed species. To some extent, the modern classification of disease within a multidimensional space (clinical, anatomoclinical, biological, genetic, radiological, etc.) strengthens this ontological conception of the nature of disease.

However, medical classification underwent a dramatic shift. As historians of medicine (Gelfand 1980; Cunningham and Williams 1992; Hannaway and LaBerge 1998) generally say, modern medicine arose with the rise of clinical medicine in the late eighteenth century (even though there are many raging controversies regarding the relative importance of certain authors, geographic center's periods: the mid-eighteenth century Edinburgh school of medicine, the nineteenth century Paris "Ecole clinique", etc. – see e.g. Ackerknecht 1967; Foucault 1963; Keel 2001). The traditional historical view states that the main focus switched from diseases as entities – as "species" or "essences" which were related to each other in a table or system that medical theory had to discover (the "nosological medicine" of early modernity) towards the diseased patient with a dysfunctional body (Jewson 1976), whose lesions or dysfunctions had to be traced back to specific symptoms. Whatever the historical value of this received view, it still implies that the importance of classification (nosography or taxonomy) decreased with the emergence of clinical medicine, anatomo-clinical medicine (namely, clinical medicine that tied the identification of diseases to the data of pathological anatomy), and then contemporary medicine, which starts with the laboratory (i.e. with biology) – as Claude Bernard accurately put it. It may be that writing a nosography was no longer the culmination of medical investigation, and that the emphasis moved towards the causal explanation of specific pathological traits and behaviors, as well as the correlated ways of curing them.

However, medical classification is still a prerequisite for any diagnosis: there is a logical priority with the question "what classes of disease exist?" to the medical clinical question "from which disease does this individual suffer?"; and nowadays, its importance is acknowledged again in many ways. As examples, first we can mention the emergence of projects on "Medical Ontology" (for example the project led by Barry Smith; see Scheuermann et al. (2009)) that intend to reformulate the extant knowledge in various fields of medicine in a simple way that would allow algorithm-aided diagnostic tools to accurately infer a diagnosis from symptoms. Second, since its inception, psychiatry has dealt in a very specific way with the problem of classifying diseases – since it has never been obvious that a specific mental suffering is a disease. And while the history of medicine can be seen as a reshaping of the boundaries within a given set of putative diseases (some diseases that were thought to be nervous have come to be seen as immunological diseases, etc.), the recent history of psychiatry at first sight presents us with a story of including and then withdrawing behaviors and mental states from the total set of mental conditions: homosexuality and fetishism left the world of psychiatry with the DSM 3; while, for example, the DSM 5 is ready to include the sexual inclination towards teenagers (hebephilia) among psychiatric diseases.

Ethics and Philosophical Issues

I mention these two examples just to stress that classifying diseases is not just a simple preliminary step to the genuinely difficult medical task of diagnosis; but is rather a crucial step in medical activity – laden with methodological difficulties, and always influenced by epistemological and philosophical ideas. Rethinking these underpinning reference points at the heart of medical thought – as well as their scientific, social, and political implications – explains why the revival of the philosophy of medicine has changed medical ethics. Even if medical ethics is not the core topic of the present book, it is hard to not point out the impact on bioethics of these new approaches in the philosophy of medicine.

Like theoretical philosophy and medicine, ethics and medicine are old companions. The Hippocratic Oath was not only about the well being of the patient, respect for private life, and the requirement of morality; it was also, and maybe mainly, about the transmission of medical knowledge and respect for the master. Whatever its purpose was, its adaptation since Antiquity into different philosophical traditions and religious contexts has preserved the Oath – with slights amendments – as a set of basic, practical, and consequences-oriented ethical guidelines. After World War II, the examination of human experimentations by Nazi physicians during the doctors' trial led to the Nuremberg Code – a landmark document in medical and research ethics. But until the early 1970s, medical ethics focused on controversial cases and moral topics that were addressed by professional organizations and religious representatives. To meet the growing need for organ transplants, an inter-disciplinary group gathered in Harvard and developed a brain-oriented definition of death. From that time, ethics committees started to change, including philosophers, historians, lawyers, social scientists, and civil representatives. Being concerned with topics including the epistemology of medicine, the concept of disease and health, causality in medicine or the positioning of psychiatry as a medical discipline (just to mention topics discussed in the present book), the evolution of the philosophy of medicine – along with the philosophy of science and biology – led to the re-evaluation of some of the basic concepts of scientifically-based medicine. That is why it is now necessary for the new philosophy of medicine to address the questions of so-called bioethics. Today, philosophical topics have direct ethical consequences,[1] and thus, medical ethics interrogations cannot be thoroughly analyzed without revisiting certain philosophical concepts.

Presentation of the Book

The current volume presents an overview of studies in the recent philosophy of medicine. The following chapters will address a set of questions that fall under the very general cartography of the philosophy of medicine outlined above. The question of health and disease and their natures – which are traditionally of high interest to philosophy because they inquire more generally into the question of normativity in life – will be addressed in several essays that also touch upon conceptual questions about the definition of medicine and its status.

The question of classification is addressed in several articles as it is central among the philosophical problems raised by medicine, especially when it comes to the field of psychiatry. Indeed, since the publication of the DSM in the 1950s, the issues of carving mental diseases into classes and justifying the attribution of one disease to one individual have remained at the core of the philosophical

---

[1] As exemplified by Singy's chapter in the present volume.

questioning of psychiatry (and in this book, given that the philosophy of psychiatry became a quite important subfield within the philosophy of medicine (Murphy 2006), we have selected an important portion of chapters that deal with psychiatry).

Finally, as highlighted above, the specificities of medical explanation have recently come under a new light, especially because of the rise of statistical methods. We did not have the ambition of addressing the question of "medical explanation" as such. The set of papers that could be ranged under the heading of questions (b) and (c) above generally address the question of how we gather, use, and assess evidence for various medical theories. Therefore, what is to be found in this book includes a sample of contributions concerning the question of evidence in medicine.

For these reasons, the book will be divided into three group of chapters that match the title: disease, classification, and evidence.

The first group of chapters addresses the problems of the nature of disease and of the status of medicine. Two chapters ask this question at the highest level of abstraction, and then three chapters question the notion of disease within specific fields of medicine.

In the first chapter, "Objectivity, Scientificity, and the Dualist Epistemology of Medicine", Thomas Cunningham tackles the long-lasting ambiguity of medicine as being both an art and a science: a science because it deals with general concepts and theories, and an art because everything rests on the contextualized singular interaction of a physician and his/her patient. For many years, the clinical sense has been understood as a kind of intuition, and good physicians are those who possess at the highest level such intuition – making medical perception the analogon of an aesthetic faculty, and medical activity a sort of art. However, Cunningham argues that this view is misleading and that there are no convincing arguments to say that medicine is an art but not a science.

In the second chapter, "The Naturalization of the Concept of Disease", Maël Lemoine addresses the very concept of disease. He challenges the philosophical project of defining the concept of disease, as was famously undertaken by Boorse among others – who defended his biostatistical theory of health as a value-free understanding of health and disease. Lemoine argues that projects concerning the conceptual analysis of health and disease are problematic since they are somehow immune to the empirical knowledge about various diseases and their mechanisms. He advocates a perspective on the nature of disease that would be founded upon the actual understanding of the mechanisms of specific diseases – rather than a priori views on normality and abnormality together with an analysis of current language.

The two following chapters by Dominic Murphy and by Steeve Demazeux are concerned with particular ranges of diseases. To begin with, their chapters are about psychiatry.

In "What Will Psychiatry Become?", Murphy questions what he calls the "medical model" in psychiatry – which means importing into the field both the knowledge of mental illness explanatory schemes (which are successful elsewhere and that operate by tracing back the psychopathological phenomena to the fundamental biological level) and our understanding of the mechanisms taking place therein. Murphy shows that such a picture does not do justice to the actual workings of psychiatry. Moreover, he argues that there is an intrinsic link between psychiatric understanding, psychology, and cognitive sciences – also arguing that the cognitive science implied here may be very different from a mere translation of the vernacular concepts used to talk about psychology (beliefs, desires, etc.).

In his chapter, "The Function Debate and the Concept of Mental Disorder", Demazeux touches upon a concept that has been crucial for the philosophy of medicine: the concept of function (since any account of disease entails an account of abnormality) and then of normality; Functional concepts indeed ipso facto normative concepts – since having a function implies the possibility of not being able to fulfill this function and therefore being abnormal. Evolutionary theory has been one of the main resources in this account of function where "functions" are not a subjective or epistemic property (i.e. functions are something that we ascribe to traits relative to our explanatory purposes and nothing more): the so-called etiological account of function, suggested by Larry Wright (1973) and then developed by Ruth Millikan and Karen Neander in the 1980s, used such a resource (Neander 1991). This indicates a way in which evolutionary theory – among all the other biological theories – provides the proper background for an investigation into the nature of disease.

Now, even if a systematic project of developing such a background has been developed under the label "Darwinian medicine" (e.g. Nesse and Williams 1996), the connections between Darwinism and medicine are scarce, and the role of Darwinian concepts, even when dealing with normativity, is not salient. However, the synthetic prospects provided by an evolutionary viewpoint could let one think that it gives us a firm standpoint to develop an understanding of what disease is – as well as its classes and types. That is why we start the next section, "Classification", with a chapter that emphasizes the role of an evolutionary perspective both in the understanding of a specific disease, and the quest to identify and classify such disease.

In this chapter, "Emerging Disease and the Evolution of Virulence: The Case of the 1918–1919 Influenza Pandemic", Pierre-Olivier Méthot and Samuel Alizon focus on a specific case – the "Spanish flu," which claimed the highest amount of lives during World War I – as a case study for the evolutionary theories involving the diffusion of infectious disease (since, for a biologist, any infection means a process of evolution of virulence). Trying to determine what the Spanish flu and its germ actually were, and what its relations are to other kinds of flu – with, in the background, a question about whether a comparable epidemic needs to be expected – Méthot and Alizon sketch the history of the recent rediscovery and sequencing of the germ responsible for the Spanish flu, and the subsequent attempts to understand its dramatic virulence within an ecological framework.

In "Power, Knowledge, and Laughter: Forensic Psychiatry and the Misuse of the *DSM*," Patrick Singy questions the courtroom use of the most common classificatory tool for psychiatrists: the DSM IV. While the main question for courtrooms concerns the capacity or incapacity of someone to refrain from some action, the DSM discusses whether someone has, or does not have, a mental condition. These two things are not logically equivalent (as the DSM's authors themselves acknowledged); some categories in the DSM such as paraphilia are forensic concepts (at least in their origin) though they seem to be medical categories. Hence, the forensic use of the DSM should be very limited. Therefore, Singy advocates a revision of these nosological categories in the DSM.

Finally, the chapter by Catherine Dekeuwer, "Defining Genetic Disease," questions the relationship between the specification of a class of diseases and the notion of genes. Some diseases have been termed "genetic diseases" – although given the nature of a gene, all pathological processes involve genetic determinants. Dekeuwer questions the legitimacy of such a concept as "genetic disease," extensively considering the case of beta-thalassemia and the policy of testing for mutated alleles and then preventing the birth of individuals who carry them. She focuses especially on the intertwining of this concept and the practice of predictive medicine.

The last section deals with issues concerning evidence in medicine. The first chapter, "Causal and Probabilistic Inferences in Diagnostic Reasoning: Historical Insight into the Contemporary Debate", by Joël Coste, puts the current practice of drawing diagnostics from a comparison of sets of data into a historical perspective. Like many disciplines following a comparable trend, medicine faces a plethoric increase in data it has to process and interpret. The multiplication of measuring devices intended to measure various biological parameters are being integrated into probabilistic models of disease. Coste approaches this situation through a historical understanding of theories and diagnostic practices.

The last two chapters deal with two current issues regarding medical judgement and medical decisions.

The chapter by Élodie Giroux, entitled "Risk Factor and Causality in Epidemiology", studies theories and concepts relative to the relationship between a pathology P and a given factor F, whether it be a determining factor or a risk factor. The central issue she addresses is: Is F the cause of P ? Or is F an element of a multifaceted cluster of factors whose members, the sub-factors, become – according to certain circumstances (ecological, social, immunological, etc.) – convergent or synergetic factors up to a point where they can be considered as genuine causes, or rather quasi-causes?

The last chapter, "Herding QATs: Quality Assessment Tools for Evidence in Medicine", by Jacob Stegenga, considers the various methods that medical literature uses to assess sets of studies concerning the same phenomena (which include randomized control trials – a subject with an extensive literature), and wonders whether some rationale can be found in the attempt to order the results of these methods in an objective and explanation-independent ranking. The chapter has a rather skeptical conclusion, claiming that there is no uncontroversial and objective way to assess sets of different tests (e.g. statistical data) concerning, for example, the efficiency of a drug.

With this book, we of course had no intention to propose another textbook for the philosophy of medicine, or to cover all the current issues discussed by philosophers and physicians alike. However, we wanted to focus on three very general topics that have been both the object of a very active philosophical discussion, and the long time centre of attention which philosophers have paid to medicine. The chapters themselves are research papers rather than synthetic and/or pedagogical presentations of an issue – or even review papers. Each one acting as a reminder of the most recent developments regarding an issue, they offer original and sometimes controversial positions. Our hope is that this sample of philosophical arguments concerning questions about health and disease, medical nosology, and medical evidence will stimulate further reflections, reading and – hopefully – contributions to the debates.[2]

Paris, France

Philippe Huneman
Gérard Lambert
Marc Silberstein

References

Ackerknecht E (1967) Medicine at the Paris hospital. Johns Hopkins Press, Baltimore

Barabasi AL, Gulbahce N, Loscalzo J (2011) Network medicine: a network-based approach to human disease. Nat Rev Genet 12:56–68

Barrett R (1995) The psychiatric team and the social definition of schizophrenia: an anthropological study of person and illness. Cambridge University Press, Cambridge

Boorse C (1975) On the distinction between disease and illness. Philos Publ Aff 5:49–68

Crichton A (1798) An inquiry into the nature and origin of mental derangement, comprehending a system of the physiology and pathology of the human mind, and a history of passions and their effects. Cadell and Davies, London

Cullen W (1784) First lines in the practice of physicks. Cadell, London

Cunningham A, Williams JP (eds) The laboratory revolution in medicine. Cambridge University press, Cambridge

Demazeux S, Singy P (2014) The DSM V in perspective: philosophical reflections on the psychiatric Babel. Springer, Dordrecht (in press)

Doll R, Hill AB (1954) The mortality of doctors in relation to their smoking habits. Br Med J 328(7455): 1529

Doll R, Peto R, Boreham J, Sutherland I (2004) Mortality in relation to smoking: 50 years' observation on male British doctors. Br Med J, 328(7455):1519

Foucault, M. (1963) Naissance de la Clinique. PUF, Paris

Gelfand T (1980) Professionalizing modern medicine. Greenwood Press, Westport

Gifford F (2011) Handbook of philosophy of science: vol 16. Philosophy of medicine. Elsevier, Amsterdam

Goldstein J (1987) Console and classify: the French psychiatric profession in the nineteenth century. Cambridge University Press, Cambridge

Guyatt A. et al. Evidence-Based Medicine Working Group (1992) Evidence-based medicine. A new approach to teaching the practice of medicine. JAMA 268(17):2420–2425

Hacking I (1975) The emergence of probability. Cambridge University Press, Cambridge

Hannaway C, La Berge A (eds) (1998) Constructing Paris medicine. Rodopi, Amsterdam/Atlanta

Howick JH (2011) The philosophy of evidence-based medicine. Wiley, Hoboken

---

[2]The author thanks Adam Hocker for careful language checking. The chapters by Joel Coste, Catherine Dekeuwer, Steeves Demazeux, and Elodie Giroux have been initially published in French in Matière Première, n° 1/2010 (nouvelle série), "Epistémologie de la médecine et de la santé", Paris: Matériologiques.

Jewson N (1976) The disappearance of the sick man from medical cosmology, 1770–1870. Sociology 10:225–244

Keel O (2001) L'avènement de la médecine clinique moderne en Europe: 1750–1815: politiques, institutions et saviors. Presses de l'Université de Montréal, Montréal

Krüger L, Gigerenzer G, Morgan MS (1987) The probabilistic revolution. 2. Ideas in science. Bradford Books

Lucas P (2001) Bayesian networks in medicine: a model-based approach to medical decision making. In: Adlassnig K-P. (ed) Proceedings of the EUNITE workshop on intelligent systems in patient care. Vienna, Austria, pp 73–97

Mayr E (1961) Cause and effect in biology. Science 134:1501–1506

Méthot PM (2011) Research traditions and evolutionary explanations in medicine. Theor Med Bioeth 32(1):75–90

Murphy D (2006) Psychiatry in the scientific image. MIT Press, Cambridge, MA

Neander K (1991) The teleological notion of 'function'. Australas J Philos 69(4):454–468

Nesse R (2001) On the difficulty of defining disease: a Darwinian perspective. Med Health Care Philos 4:37–46

Nesse RM, Williams G, (1996) Why we get sick: the new science of Darwinian medicine. Vintage Books, New York

Nikovski D (2000) Constructing Bayesian networks for medical diagnosis from incomplete and partially correct statistics. IEEE Trans Knowl Data Eng 12:4

Pauling L, Itano HA, Singers SJ, Wells IC (1949) Sickle-cell anaemia, a molecular disease. Science 110:543–548

Pinel P (1798 [1813]) (6th ed.). Nosographie Philosophique. Richard, Caille, et Ravier, Paris

Russo R, Williamson J (2007) Interpreting causality in the health sciences. Int Stud Philos Sci 21(2):157–170

Sadegh-Zadeh K (2012) Handbook of analytic philosophy of medicine. Springer, Dordrecht

Sauvages F Boissier de (1772) Nosologie méthodique, ou distribution des maladies en classes, genres et espèces selon l'esprit de Sydenham et la méthode botanique. Mercier, Lyon

Scheuermann RH, Ceusters W, Smith B (2009) Toward an ontological treatment of disease and diagnosis. In: Proceedings of the 2009 AMIA summit on translational bioinformatics, pp 116–120

Scull A (1975) From madness to mental illness: medical men as moral entrepreneurs. Archives européennes de sociologie 16:218–251

Smith R (2002) In search of 'Non-Disease'. Br Med J 324:883–885

Spirtes P, Glymour C, Scheines R (2000) Causation, prediction and search. MIT Press, Cambridge, MA

Wachbroit R (1994) Normality as a biological concept. Philos Sci 61:579–591

Wakefield JC (1992) The concept of mental disorder. Am Psychol 47:373–388

Whytt J (1765) Observations on the nature, cause, and cure of those disorders commonly called nervous, hypochondriac or hysteric. Hamilton, Edinburgh

Williams GC, Nesse W (1991) The dawn of Darwinian medicine. Q Rev Biol 66(1):1–22

Wright L (1973) Functions. Philos Rev 82:139–168

Objectivity, Scientificity, and the Dualist Epistemology of Medicine

Thomas V. Cunningham

Abstract This paper considers the view that medicine is both "science" and "art." It is argued that on this view certain clinical knowledge – of patients' histories, values, and preferences, and how to integrate them in decision-making – cannot be scientific knowledge. However, by drawing on recent work in philosophy of science it is argued that progress in gaining such knowledge has been achieved by the accumulation of what should be understood as "scientific" knowledge. I claim there are varying degrees of objectivity pertaining to various aspects of clinical medicine. Hence, what is often understood as constituting the "art" of medicine is amenable to objective methods of inquiry, and so, may be understood as "science". As a result, I conclude that rather than endorse the popular philosophical distinction between the art and science of medicine, in the future a unified, multifaceted epistemology of medicine should be developed to replace it.

Introduction

In philosophy, clinical medicine is commonly said to have a dualistic nature, to be both science and art.[1] How this assumption is interpreted is important because the extent to which we view medicine as science rather than art affects our epistemological expectations of medicine. For example, if we hold that medicine is a science, it has been argued we should thus only expect it to meet scientific standards of inquiry, namely, the acquisition of objective knowledge. On such reasoning, medicine need not meet additional moral standards of inquiry, such as being sensitive to

---

[1] I wish to thank Philippe Huneman for helpful comments on a draft of this paper.

T.V. Cunningham (✉)
Division of Medical Humanities, University of Arkansas for Medical Sciences,
4301 W. Markham St., #646, Little Rock, AR 72205, USA
e-mail: tcunningham@uams.edu

P. Huneman et al. (eds.), *Classification, Disease and Evidence*, History,
Philosophy and Theory of the Life Sciences 7, DOI 10.1007/978-94-017-8887-8_1,
© Springer Science+Business Media Dordrecht 2015

patients' health care needs and how they are met (Munson 1981). But of course, medicine without moral sensitivity would be deeply flawed, as it would forsake a basic aim of benefitting the patient through restoration and healing. Hence, so this reasoning goes, we should be persuaded to adopt a *dualist epistemology of medicine*; we should recognize two equally fundamental ways of medical knowing: in terms of objective scientific knowledge of biology and physiology, and subjective personal knowledge of the craft of patient care.

While I accept that ethical medicine must be sensitive to patients' health care needs and how they are met, the claim that this belief provides a reason to adopt a dualist epistemology of medicine is not persuasive. Indeed, I contend this doctrine has pernicious affects on our understanding of integral aspects of clinical medicine, because accepting it implies that certain clinical knowledge – of patients' histories, values, and preferences, and how to integrate them in decision-making – cannot be scientific knowledge. Yet, decades of work in clinical decision science suggests this knowledge is already being attained and used, altering how clinicians provide care (e.g., Weinstein and Fineberg 1980; Ende et al. 1989; Deber et al. 1996; Stiggelbout and Kiebert 1997; Levinson et al. 2005). If we aim to accurately capture the epistemic structure of medicine, including types of knowledge commonly relegated to the undifferentiated heap of the "art" of medicine, then this aim motivates a reassessment and challenge of the dualist epistemology of medicine.

Moreover, recent work in history and philosophy of science suggests that the art/science distinction rests on deeply flawed and hackneyed assumptions about science, as value free inquiry (*e.g.*, Longino 1990; Proctor 1991; Dupré 1993; Nelson and Nelson 1996; Lacey 1999; Douglas 2009). Thus, the arguments given here against a dualist epistemology of medicine also find a second motivation, of questioning a common thesis in philosophy of medicine in light of recent progress in philosophy of science.

The plan of the paper is as follows. It first reconsiders a classic debate over the scientificity of medicine, which shows that the vision of science assumed for juxta-position with clinical medicine underpins conclusions about the scientificity of medicine. That is, whether we see medicine as a science rather than an art will depend chiefly on the extent to which we believe medicine is inherently "subjective" and "value-laden" versus "objective" and "value-free," and the extent to which science is not. Drawing on recent work in history and philosophy of science on the conceptual complexity of objectivity and subjectivity (Douglas 2004, 2009), the paper next argues that a dualist epistemology of medicine assumes an antiquated dichotomy between pure objectivity and pure subjectivity, where science aims at (and achieves) the former and anything that does not is not science. If we reject this dichotomy, as it is argued we should, then what is important is no longer whether medicine is a science, but the extent to which aspects of clinical medicine may be said to be objective, and therefore, amenable to scientific methods of inquiry. As two brief case studies show, while there remains (and will always remain) a degree of subjectivity in clinical medicine, this does not entail that it cannot be a science, once science is understood as admitting of multiple types of objectivity and as incorporating values.

Distinctions in the Art/Science Debate

Being a Science Versus Being Scientific

Over 30 years ago, Lee Forstrom argued clinical medicine is not only scientific, but also is an autonomous science. Following Braithwhite, Forstrom defined a science in terms of two criteria, whether it has its own natural domain of inquiry and whether it aims at establishing general laws explaining the phenomena of that domain (1977, 8–9). Rendered in light of contemporary concepts in philosophy of science, we may interpret Forstrom as arguing that medicine has both a unique domain of inquiry and that it aims at robust generalizations.

According to Forstrom, the domain of clinical medicine is the living human being, which is both its object of inquiry and "its usual experimental context" (15). Yet, as human illness manifests across levels of analysis, from molecules and organs to organ systems and social systems, the clinician "must interpret and evaluate the effects on the organism of social and economic as well as physical and biologic factors" (9). Thus, medicine's unique domain is the sum total of levels of analysis required to understand health and disease in a living, embodied person. It is not simply an aggregate of the other sciences that explain phenomena in these domains, such as molecular biology, genetics, physiology, psychology, and economics, because medical science synthesizes the results of these domains for the purpose of developing knowledge designed for individual patient care.[2] Medicine is thus directed at knowledge about patient care rather than about biopsychosocial phenomena isolated from the context of human well being and suffering.

In response to Forstrom, Ronald Munson argues medicine is not, and will never be, a science, even though it is *scientific*. Using Forstrom's criteria, Munson argues that despite the fact that the notion of a unique domain of inquiry is vague, medicine nevertheless fails to have one. Because, he says, simply identifying a concern with the health and disease of living humans, and a requirement that this concern be expressed in considerations of many levels of analysis, fails to demarcate medicine from other fields, such as "medical sociology, epidemiology, bacteriology, biochemistry, and social work" (1981, 186). Moreover, distinguishing medicine from these fields by appealing to medical intervention as the defining aspect of clinical medicine will not do, because that would be patently circular.

---

[2] This depiction accords well with Engel's "biopsychosocial model of medicine" (Engel 1977) and the more recent model of "patient-centered medicine" (Bardes 2012).

What is at issue here is the type of generalizations clinical medicine aims at and how robust they are. To see this, notice that Munson's main objection is that not only is medicine not a science, but also, it can *never* be a science. While Munson recognizes that medicine is scientific, he rejects the claim that it is a science because of how he defines science. Munson holds something is a science if and only if it aims to generate robust generalizations; thus, the basic aim of science "is the acquisition of knowledge and understanding of the world and things that are in it," (190), no more and no less. For a scientist to justify her work, she "need only demonstrate that it is likely to increase our knowledge" (191). For a physician, however, solely appealing to increasing knowledge is insufficient and actually negligent. Since the aim of medicine is "to promote the health of people through prevention or treatment of disease," to justify her work, "the medical researcher must, in effect, present a dual justification: (1) the work will increase our knowledge; (2) the knowledge will be relevant to the aim of medicine" (*ibid.*).

Munson's response is perhaps the clearest of many attempts in the past three decades to justify a common view, that medicine is both science and art. That is, on the one hand, it aims at robust generalizations, while on the other hand it aims at idiosyncratic inferences concerning the treatment of particular persons. Because of these dual aims, medicine is bound to be concerned with patients' assessments of health, which entails a consideration of patients' values. Consequently, Munson concludes medicine has an inherently subjective, moral component, whereas science lacks such a component because of its function, to generate pure, objective knowledge.

Values, Scientificity, and Objectivity

Beneath Munson and Forstrom's debate lie assumptions about what characteristics must be present in order for science to aim at robust generalizations. Specifically, this debate shows that what justifies construing medicine in terms of a dualism between science and art is another assumed dualism, between inquiries that are "value-free" and those that are "value-laden," where the former pertain to the science of medicine and the latter to its art. For Munson, understanding what it means to aim at robust generalizations requires conceiving of them in terms of objective, value-free knowledge of the world. These are the targets of science, whereas medicine aims *also* at a subjective understanding of the patient. However, by questioning this second-order dualism, we can show that there are better ways to understand "science," and thus, better ways to describe the sense in which science and medicine aim at robust generalizations.

Consider the approach adopted by Gorovitz and MacIntyre in a classic paper from the same era. Science, they say, does not only aim at universal knowledge of properties, kinds, and generalizations linking one to the other; it also aims at generalizations about particulars. And, medicine is a science, so understood. For the clinician, understanding what makes a particular individual distinctive is paramount, even if this understanding comports poorly with medical theory. Whereas a scientist aims to yield abstract generalizations from his or her experiments, rather than fuller knowledge of the specific features of samples being studied, for the clinician working with particular patients, "how such particulars differ from one another in their diversity thus becomes as important as the characteristics they commonly share" (1976, 59).

Gorovitz and MacIntyre's claim that medicine is a science hinges on their rejection of the fact-value dichotomy, which they say gives a false impression of the epistemology of science. It is the familiar thesis that sciences generate statements of fact, which cannot entail statements of value, that they contend leads to the erroneous view that natural sciences are not concerned with particulars, and as medicine is clearly so concerned, that medicine is not a science.[3] For Gorovitz and MacIntyre, then, medicine is a science, and that it is so is entailed by an account of scientificity that differs from Forstrum's and Munson's. Sciences *are* concerned with understanding particular phenomena, such as particular hurricanes, tsunamis, election results, and stroke victims. Hence, the fact that medical theory and practice are focused on understanding particular patients does not imply medicine is not a science.

Taken together, the claims made in these classic papers indicate that there are at least three different concerns at issue in debates about the dualist epistemology of medicine, each of which can be simply rendered in terms of a second-order dualism or distinction. One concern is with subjectivity and objectivity, specifically as manifested in a dualism between subjective and objective knowledge. Another is with value-free versus value-laden types of inquiry, and their relation to the production of knowledge. A third concern is captured in the distinction between general explanations and explanations of particulars.

Given that each of these three distinctions admits of its own literature, it would be foolish to attempt to give a full characterization of any of them here.[4] My aim is more modest, namely to show how attending to the assumptions one holds regarding each of them supports ones epistemology of medicine, and moreover, that certain (more tenable) assumptions suggest that a multifaceted epistemology of medicine is warranted, rather than a dualist one distinguishing simply between science and art.

---

[3] As an aside, this claim warrants comment. It is not clear that ethical non-naturalists need be troubled by Gorovitz and MacIntyre's assertion here. They need only deny that factual information is sufficient for informing claims about what is good, not that it can play a (non-sufficient) warranted role in justifying inferences about what is good for a patient or other agent in the health care system.

[4] Indeed, for example, the issue of generality in explanation has been with us since the Ancients. Ancient Greek thinkers also distinguished between *episteme* and *techne*, a distinction based in part on the claim that the best explanations are those that are timeless and apply with broad generality. However, though early Greek thinkers also distinguish between these forms of knowing, as discussed below (n. 8), these distinctions do not match the contemporary distinction between art and science well as it is described here. See Parry (2009) for a detailed review of the diversity of Ancient Greek views on this topic and the many ways they relate to current epistemology.

When understood in terms of objectivity and subjectivity, the debate over whether medicine is a science comes down to whether medicine is "purely objective" and aims at the accumulation of objective knowledge, or whether it includes an inherently "subjective" component. This "subjective" component has been rendered in terms of personal values in the debate over the scientificity of medicine. In this way, we see the interplay between the value-free/value-laden distinction and the distinction between objectivity and subjectivity, in that medicine is an art if it aims at understanding patients' subjective knowledge of illness in terms that are patently laden with the patient's values. Likewise, medicine is understood as a science in as much as it aims to understand patients' diseases in objective terms, meaning those that are disconnected from the values of particular patients and clinicians.[5]

Distinguishing between general explanations and explanations of particulars also relates to the other two distinctions. If understood as a science, medicine is taken to aim at knowledge that holds of patients in general, indeed *because* it aims at objective knowledge, free from the values of particular patients and clinicians. And, medicine is art insofar as clinicians aim to skillfully bring these generalizations to bear on subjectively understood, value-laden illness in particular instantiations; that is, in particular patients.

Eric Cassell, a longstanding proponent of the dualist epistemology of medicine (*e.g.*, 1995, 2004), provides a paradigmatic example of how these distinctions interrelate in philosophical explorations of medicine. Cassell argues that in practice physicians adopt a narrow understanding of the concept of objectivity and a multifaceted understanding of subjectivity. Imagine you feel feverish, he says. You are achy and have cold sweats. You feel ill. If you go to a physician and she takes your temperature, then, "the reading on the clinical thermometer is an objective measurement of an elevation of body temperature. The feeling of feverishness is subjective because a feeling can only be experienced by the subject" (Cassell 2004, 171). This is one sense of what it means to be subjective; it is to feel a certain way, which can only be felt by you, the subject. There is also another sense, which is associated with your ideas *about* the way you feel. You may think that your feelings of achiness warrant the belief that you have a fever. According to Cassell, that idea is subjective in a second sense. Thus, on this view, how you feel and what you reason about your state of affairs in light of your feelings are both subjective. But, there is also a third sense of 'subjective' in medicine: "your *statement* that you feel feverish is also considered subjective...What the words *mean* is not something outside observers can hold in common," hence, they are subjective too (*ibid.*; italics in original).

Notice here that for Cassell, being subjective connotes being specific, local, and particular. Individual persons have particular feelings, ideas, or understandings of meaning. However, being objective is associated with generality: a thermometer reading is taken to be objective by contrast to being felt solely by the subject – it is

---

[5] In his *The Wounded Storyteller*, Arthur Frank (1995) develops an account of illness as subjective experience and disease as the objective description of that subjective experience in biomedical terms. It is in this sense that I use terms such as "illness" and "disease."

valid everywhere, no matter who wields the apparatus, as long as it is used correctly.[6] Also, being objective is associated with being general in the sense that there is general agreement about objective features of the world, in contrast to the particular meaning of statements as understood by specific persons.

Thus, underlying debates about the scientificity of medicine are assumptions about the meaning of objectivity, which is intimately related to the role of values in, and generality of, the target knowledge of interest. It is assumed that the clinician is tasked with acquiring two types of knowledge about the patient, objective (scientific) knowledge, for which there are general, measurable facts of the matter, and subjective knowledge, for which there are particular, incorrigible idiosyncrasies and thus, no facts of the matter.[7]

Objectivity and the Scientificity of Medicine

Now, one may reasonably wonder whether it matters that some philosophers defend a dualist epistemology of medicine. There are at least two reasons to think that it does. First, if we accept a dualist epistemology of medicine – as inherently both "art" and "science", both "objective" and "subjective," both "value-free" and "value-laden" – then such common activities as a clinician inquiring about a patient's symptoms, beliefs about the genesis of his complaint, or way of speaking about his illness, become activities that cannot be objectively characterized. That is, if medicine is both science and art, then we must agree with Cassell that "establishing a scientific basis for dealing with values and human qualities" is "doomed…Instead, each physician must solve the problem internally" (2004, 19–20).[8] Second, another

---

[6] In contrast to Cassell's assertion, Hasok Chang's (2004) work on the science of thermometry shows that the standardization of the activity of measuring "temperature" over hundreds of years is what makes this example appear as an innocuous instance of the elucidation of a objective fact about a patient. However, Chang's account of the evolution of the concept of temperature shows that such facts require literally centuries of research and debate in order for the idiosyncrasies of experimentation to be codified into a broadly accepted physical theory of temperature measurement.

[7] Another context in philosophy of medicine where the relationship between objective and subjective knowledge figures largely is debates over the meaning of the concepts, health, disease, and illness. Beginning with Boorse's account (1977, 1997), some argue that health has meaning by contrast with disease, which is best described in objective, "biostatistical" terms, or in terms of species typical functioning. Yet, others argue that these foundational medical concepts are thoroughly subjective due to the normative, evaluational aspects of medical reasoning and nosology (*e.g.*, Nordenfelt 1987). And, yet others contend that concepts like health and disease are normative *and* objective, proposing a hybrid account of sorts (Lennox 1995; Schaffner 1999). Finally, others argue that understanding these concepts philosophically is a quixotic pursuit, with no bearing on medical practice (Hesslow 1993). Taking a stance on this literature lies beyond the scope of this inquiry.

[8] This too is a problem that extends historically to the Ancients. As noted (n. 4), Ancient Greek philosophers distinguished between different ways of knowing, including *episteme* and *techne*. However, different thinkers interpreted these terms quite differently. For example, in the

reason that the dualist epistemology matters is that it is common in philosophy of medicine (*e.g.*, Waymack 2009; Saunders 2000; Cassell 1995; Malterud 1995; Battista et al. 1994), which threatens to distance work in this field from important progress elsewhere in philosophy, especially in philosophy of science.[9] That is, given the progress made in recent years on the question of whether science is value-laden or 'purely objective' in philosophy of science, if philosophy of medicine ignores this work it adopts an antiquated theory of science, which threatens to render it obsolete.

The Irreducible Complexity of Objectivity

One way to challenge the dualist epistemology of medicine is to challenge its conceptualization of scientific objectivity. In light of recent work in history and philosophy of science, objectivity may be seen as far more complex than discussants in the art/science debate suppose.[10] Consequently, the notion of "value-free" aspects of clinical medicine is a nonstarter. Therefore, clinical medicine should be understood as an integrative science that draws on various methods, which are objective by varying degrees.

In its contemporary usage, the concept of objectivity is, as historians of the notion have put it, "hopelessly but interestingly confused" (Daston and Galison 1992, 82). Following Heather Douglas (2004, 2009), we may distinguish between different senses of objectivity implicit in the broader concept by attending to the different ways objective claims are *produced*. Douglas distinguishes three categories of processes that result in objective claims: interactions with the world (such as experimentation or observation), individual thought processes (particularly reasoning leading to certain claims), and social processes for generating claims (such as polling, voting, or collaboration).

As illustrated by Cassell above, from the clinician's perspective, interacting with patients may be seen as an instance of an interaction with worldly phenomena.

Nicomachean Ethics (especially *Book VI*), Aristotle describes these two types of knowledge as more general, in contrast to a third type of knowledge of how to act rightly in particular contexts, known as practical wisdom or *phronesis* (Aristotle 2000). It is fascinating that Ancient Greek thinkers took medicine, along with navigation, as an exemplar of practices where all types of knowledge were required (Jaeger 1957). Although these discussions are clearly relevant to modern debates about the epistemology of medicine, contemporary scholars are in agreement that the Ancient Greek conceptions of knowledge do not mirror our own understanding of art as a craft and science as objective facts (Hofmann 2003; Evans 2006).

[9] The same might be said for empirical work in applied ethics, however, for the sake of brevity that point will not be made here.

[10] This argument could be expanded to draw on the considerable philosophical and historical literatures on objectivity and science (e.g. Nagel 1979; Longino 1990; Proctor 1991), but doing so is outside the scope of the present discussion.

Though a patient is a person, he is also a phenomenon to be studied, to be poked and prodded, in order to generate evidence for knowledge claims. To make such claims, physicians procure evidence through multiple avenues, such as different types of diagnostic tests (e.g., genetic, blood, and imaging), and inquire whether the evidence supports inferences about the patient's illness. On Douglas' typology, this is *convergent* objectivity, where convergence of sufficiently independent lines of inquiry yields "increasing confidence in the reliability of the result" (2009, 119–120).

Interacting with patients may also be understood as a social process, for instance, of eliciting information about the patient's illness, of healing, or of deliberating about treatment options. These processes may also be understood as generating objective claims. *Concordant* objectivity occurs when "some set of competent observers all concur on [a] particular observation" (126). *Interactive* objectivity denotes moments where persons deliberate "to ferret out the sources of their disagreements" before certifying a claim (127). In the clinical context, concordant objectivity may be exemplified by physician consultations or second opinions. In each case, the question is whether multiple observers will agree on a patient's diagnosis, prognosis, and treatment options; if so, then in this sense the agreement conveys that these are objective claims about the patient. Interactive objectivity is exemplified by treatment decision-making and team-based approaches to clinical care, where in both instances persons deliberate over whether a choice is correct in light of what is known about a patient.[11] According to Douglas, the more diverse the deliberators and the more robust the disagreement and deliberation, the more objective this type of objectivity will be.

Individual thought processes could also be described as objective. In one sense, to be objective is to think about phenomena while keeping personal 'distance' from it. That is, *detached* objectivity follows from a "prohibition against using values in place of evidence" (120); the investigator is prohibited from appealing to her values in making inferences about the happenings of the world. This seems to be the kind of objectivity intended by Munson in his characterization of science, where scientists aim at producing general knowledge, and nothing more. Yet, Munson's characterization of science is ambiguous in that it also implies *value-free* objectivity, which is more restrictive than detached objectivity, because it denotes a process where all values are prohibited from entering into reasoning. If science is characterized as lacking an inherent moral principle, as Munson holds, then this suggests values are banned from scientific reasoning, which is a stronger prohibition than that they cannot serve as components of inferences (detached objectivity) or that one must adopt a neutral position with regard to the values at play in inquiry (*value-neutral* objectivity).

---

[11] For a lively, careful discussion of the philosophical implications of team-based care, see the contributions to King et al. 1988.

Scientificity and the Epistemology of Medicine

Whether clinical medicine is both art and science depends on how one defines "science". In the art/science debate, to be a science is to be "value-free," "objective," and to aim at (robust} generalizations. But as Douglas (2004) argues, the meaning of "objectivity" is irreducibly complex; consequently, the extent to which being value-free is a hallmark of science is an open question that depends for its answer on the extent to which science exhibits various types of objectivity. Thus, if science is not value-free in the requisite sense – of value-free objectivity defined above – then the claim that medicine is not a science becomes unsupportable. Just as other sciences exhibit types, and hence degrees of objectivity, so too does medicine. Accordingly, just like other sciences, medicine may be seen as a science despite the fact that it is not "value-free."

There are good reasons to think that value-free objectivity is not and should not be a hallmark of scientific inquiry. As Douglas argues, scientists routinely make decisions about research based on various methodological and ethical values. Scientists also dispute the relative importance of different epistemic values and their implications for hypothesis acceptance. Furthermore, the distinction between epistemic and non-epistemic values is dubious. Finally, scientists have a responsibility to consider the consequences of errors in their reasoning. What follows from this is that the role values play in science indicates that the value-free ideal of objectivity is also a nonstarter. Values are ever-present in science; understanding the roles they play in inquiry and the extent to which they are justified is what is important.

Values play many roles in medical reasoning. Hence, a satisfactory epistemology of medicine should not be dualistic, but should be both unified and multifaceted. It should be possible to describe the moments where, for example, detached objectivity is warranted or inapt, or where convergent objectivity justifies a claim that is nonetheless challenged through processes described by concordant objectivity. To put it another way, if we shift from a dualistic epistemology of medicine to a unified and multifaceted one, we may draw upon rich philosophical accounts of the multi-level nature of explanation in medicine (Schaffner 1993) in order to justify the types and degrees of objectivity operative at each level and the extent to which they interact in the making of justified medical claims. On such an account, clinical medicine is a science through and through, only to be a science is no longer to be "objective" in a simple sense of being value-free; rather, to be objective is to be produced by a process one can rely on, a process that is likely to be trustworthy.

However, though we may be better positioned to evaluate the implications of the art/science distinction in medicine by considering recent work in history and philosophy of science, we may nevertheless still believe the dualist epistemology has its virtues. Principle among these might be its emphasis on the distinction between general and particular knowledge claims, an area of inquiry where history and philosophy of science has made far less recent progress than in the understanding of values and objectivity. That is, though we may follow Douglas and others in shifting an emphasis to how knowledge claims are produced to understand the roles values play in them and the senses in which they are objective, it is not clear that this is helpful for characterizing the extent to which these claims are more or less general or particular, and what this means for their validity, reliability, or meaningfulness.

Consider that we may speak of a "myocardial infarction" as a type of event or as a token event. It is not clear whether clinicians who use this language – or language of "swollen," "sharp pain," or "anxiety" – generally mean to invoke just the type or just the token event. Which, or whether they are being ambiguous, will be a matter of the pragmatics of medical practice, and is not something that can be decided in the current inquiry. Moreover, it is also unclear how clarifying the multiplicity of ways science is value-laden and the complexity with which it aims at objectivity will aid in characterizing those pragmatics, though I assume in time they will.

Consequently, if these remarks about the complexity of objectivity and the role played by an antiquated concept of objectivity in the dualist epistemology of medicine are cogent, then they suggest at most that the art/science distinction rests on shaky ground. If, as has been argued, science is value-laden, then the mere fact that what is often called the "art" of medicine requires eliciting patients' values does not entail medicine cannot, therefore, be a science. With objectivity so understood, the traditional art/science distinction may thus reduce to an ancient, and perhaps intractable puzzle about the relationship between general and particular knowledge claims.

A Role for a Unified Epistemology of Medicine: Two Case Studies

Absent sound philosophical reasons for adopting a dualist epistemology of medicine, I contend it should be rejected because of its pernicious effects, which I describe in this section by considering examples from recent work on decision-making in hereditary breast and ovarian cancer syndrome (HBOC) and end-of-life care. In both cases, one finds many aspects of clinical medicine that are routinely understood under the rubric of the "art" of medicine, but which are better understood when depicted as part of the "science" of medicine, because doing so allows for the assimilation of this research into the domain of unified medical knowledge.

Pathophysiology, Psychology, and Social Science in Hereditary Breast and Ovarian Cancer

HBOC is defined in terms of a known genetic predisposition to breast and ovarian cancer. Many factors must be considered in its diagnosis, but the determining one is returning a positive result for mutations in the *BRCA1* or *BRCA2* genes (Rubenstein 2001). In order to qualify for a genetic test, a patient must meet certain criteria, including having a first degree relative with a known mutation, being of Ashkenazi Jewish descent, or receiving a diagnosis of breast cancer before age 45 (National Cancer Institute 2011). If a patient is diagnosed with HBOC, this licenses a number of inferences about processes that are occurring in her cells, depending upon the mutation she harbors (Turner et al. 2004). While much is known about the genetics and physiology of this syndrome, the study of HBOC is still in its infancy, so it is known with varying degrees of uncertainty. Despite this uncertainty, these aspects of the clinical science of HBOC surely fall under the rubric of the "science" of medicine on any account.

However, we know much more about HBOC than simply its pathophysiology. We also know how the ways in which clinicians interact with patients may affect their decision-making. And, we know what patients' typical emotional reactions will be when faced with the prospects of having HBOC. Appreciating this research, described below, on various phenomena arising from typical clinical encounters in HBOC suggests that what is often understood as the "art" of medicine is also a science, though in the psychological and social sciences. It aims to measure qualities of particular social beings and social relations. And it studies agents who seek care, their loved ones, the professionals who provide care, and the relationships among them. Through increasing success at such measurement, increasing development and application of statistical techniques, increasing conceptual progress, and increasing innovation in experimental design, we are learning about these relationships in ways that support interventions upon them. Thus, the art and craft of medicine is constituted by diverse studies of social relations in medical practice and their application to particular moments of patient care.

Empirical studies of the psychosocial aspects of HBOC have resulted in a rich portrait of what it means to face an HBOC diagnosis, how patients and family members make treatment decisions, and what the consequences of their choices commonly are. For example, we know that genetic counselors are far more disposed to choose genetic testing and prophylactic surgery than their patients (Matloff et al. 2000). And we know that what is most important to patients who face decisions about testing and surgery is information about their test results and their family history. Yet, also of importance are concerns about the risks of surgery, the timing of interventions in their lives, and the impact treatment will have on sexuality (Ray et al. 2005). Finally, for those who choose testing, we know that irrespective of their test results, patients will feel a mixture of sadness, anger, guilt, and relief; and many will worry about insurance discrimination (Lynch et al. 1997).

Though this description of HBOC is abstract and simplified, it suffices to illustrate both how a complex understanding of objectivity is useful for characterizing the scientificity of medicine and why it is better to understand medical knowledge as scientific, rather than as both science and art.

The principle justification for a dualist epistemology of medicine resides in the belief that there are certain aspects of the craft of medicine that are inherently subjective and particular, meaning they are value-laden, and hence, inaccessible to scientific methods of inquiry. The examples of such aspects given above by Cassell are the values of the patient, the idiosyncrasies of clinical judgment, and the emotional influence on patient and physician cognition during all aspects of clinical interactions. The position argued for here is that these features may also be understood as being objective, once a simplistic account of objectivity is identified, challenged, and replaced with a more nuanced account. On this view, empirical studies of phenomena like clinicians' biases and patients' emotional responses to various moments in treatment provide knowledge that is objective, and in an important way, in the same sense as knowledge of the molecular processes that cause cancer. Both types of knowledge are the result of many processes of data collection and inference. These processes will be objective to varying degrees, if modeled in terms of the types of objectivity above. Whereas our knowledge of the molecular pathophysiology of HBOC may be a product of processes where concordant, convergent, and detached objectivity are more salient than other types, it is also true that our knowledge of the psychosocial aspects of HBOC are produced by processes where concordant, interactive, and value-neutral objectivity play prominent roles. Hence, it is not the case that what has been characterized as the art of medicine is incorrigible by appeal to scientific inquiry; rather, it is, and this entails that there may be a science of the art of medicine. Furthermore, just as objective knowledge of pathophysiology is necessary for the optimal delivery of patient care, so too is objective knowledge of the psychosocial aspects of medicine instrumental for optimal care.

End-of-Life Care in the Intensive Care Unit and the Scientificity of Medicine

Research on decision-making in end-of-life care is another, and perhaps better, case where important recent progress has been made in scientifically studying aspects of care that would traditionally be confined to the "art" of medicine. For many people, life will end in an institutional setting; indeed, recent studies showed that for over 65 % of subjects life ended in an institutional setting, including a hospital or nursing home; and, for those who died at home, over 60 % received some type of nursing or hospice care at home before death (Teno et al. 2004). In recent years, clinicians, social scientists, and applied ethicists have endeavored to describe how people die in institutional settings and to suggest ways for optimizing these most common ways of ending life (*e.g.*, Kaufman 2005). What considering a tiny but representative portion of this literature indicates is that, like in the case of HBOC, studies have developed significant, empirically justified knowledge describing end-of-life care from the perspectives of providers, patients, caregivers, and other stakeholders and decision makers. In the context of the current argument, this research provides further evidence for the view that a dualist epistemology of medicine has pernicious effects: specifically, if one accepts it, then the knowledge gained by such studies must be dislocated from other medical knowledge that is equally important to providing quality end-of-life care, namely, that which justifies scientific inferences about a patient's prognosis. With such dislocation there is no coherent way to describe how to take into account both how patients understand dying in institutional settings or what it means to provide quality end-of-life care, and how to provide quality medical interventions at the end of life. And this is deeply problematic because integrating these different types of knowledge is required in end-of-life care, as it is in all medical practice. Yet, if one adopts the science/art distinction, one should expect that such an integration is not only hopeless, but also impossible, because on that view to try to assimilate art into science or *vice versa* is to make a category error.

In a recent review, physician J. R. Curtis argues that "[Intensive Care Unit] clinicians should approach the family conference with the same care and planning that they approach other ICU procedures" (Curtis et al. 2001). That is, he believes it is just as important to understand how to communicate well about end-of-life decisions, as it is to understand how to perform a procedure such as a tracheostomy. Moreover, knowledge of how to do both well is not just equally important, it also may require the same types of reasoning.

As Curtis has shown in many subsequent studies, to be a better communicator requires developing an expertise in understanding how physicians communicate poorly and how they communicate well. For instance, in a recent paper, he and his colleagues show that there are four distinct roles that physicians take on when discussing surrogate decision-making regarding life support decisions. Most physicians adopt a collaborative role, defined as providing medical information, eliciting patient's values, and making treatment recommendations. However, others adopt what they describe as "directive," "facilitative," or "informative" roles in the decision-making process (White et al. 2010). If philosophers are to adequately characterize clinical research such as this, then the dualist epistemology of medicine must be rejected and superseded by a more cogent account.

Consider other recent work on end-of-life care. A recent study of the psychology of clinical decision-making in the ICU shows that physicians' beliefs about the appropriateness of withdrawing life support strongly correlate with whether patients in the ICU receive the option to withdraw treatment (Schenker et al. 2012). This research suggests that by better understanding the mental mechanisms by which physicians form beliefs it may be possible in the future to create interventions to increase the quality of end-of-life care, in terms of increasing the goodness of fit between presented treatment options and patients' intuitions about quality of life. Consequently, the logic of this research presupposes that by empirically studying the "art" of medical practice using common scientific methodologies, it will be possible to both better understand clinical practice and create empirically derived interventions for bettering patient care.

As in the case of HBOC, careful attention to research studying various aspects of clinical practice and decision-making in end-of-life care indicates that there are many instances where scientific methods are fruitfully applied to patient care. What results are measures of various aspects of patient care that have the promise of revolutionizing practice once better understood. Such measures are out of place if one joins scholars like Cassell in assuming that there is an art to medicine that cannot be studied scientifically, and for which no progress can be made other than by the apprenticeship model. Given the value of these measures, it is only reasonable to conclude that the time has come to move past the dualist epistemology of medicine and to begin the process of crafting a new, coherent epistemology that is multifaceted while also remaining unified in recognizing the persistent scientificity of medical theory and practice.

Conclusion

This paper argues a dualist epistemology of medicine has significant and pernicious implications. It implies that certain clinical knowledge – such as, of patients' histories, values, and preferences, and how to integrate them in decision-making – cannot be scientific knowledge. Moreover, the distinction between an "art" and "science" of medicine rests on flawed and antiquated conceptions of science, as characterized above. By considering recent progress on the question of whether science is value-free, and relatedly, the conceptual complexity of objectivity and subjectivity, it has been argued there may be varying degrees of objectivity pertaining to various aspects of clinical medicine. Hence, what is often understood as constituting the "art" of medicine is also amenable to objective methods of inquiry, and so, may be understood as "science". Therefore, the popular philosophical distinction between the art and science of medicine ought to be rejected and in its place a unified, multifaceted epistemology of medicine should be developed.

The upshot of rejecting a dualist epistemology of medicine is that it allows one to make explicit and to critically evaluate the role of values in medical science. It stands to reason that different aspects of medicine, such as the pathophysiological and psychosocial, will have very different degrees of objectivity, correlative with the different roles values play in them. But, these should be understood as differences of degree, not in kind. Thus, it is better to see medicine as an integrative science aiming at multi-level explanation in the service of patient health, rather than as a science on the one hand and an art on the other. What remains an open question, however, is whether issues arising from the generality and particularity of knowledge claims in medicine continue to be salient in light of new understandings of the complexity of objectivity and roles of values in science. This is but one of the important issues facing those who aim toward a unified epistemology of medicine.

References

Aristotle (2000) Nicomachean ethics (trans: Crisp R). Cambridge University Press, New York
Bardes C (2012) Defining "patient-centered medicine". N Engl J Med 366:782–783
Battista RN, Hodge MJ, Vineis P (1994) Medicine, practice and guidelines, the uneasy juncture of science and art. J Clin Epidemiol 48:875–880
Boorse C (1977) Health as a theoretical concept. Philos Sci 44:542–573
Boorse C (1997) A rebuttal on health. In: Humber J, Almeder R (eds) What is disease? Humana Press, Totowa, pp 3–134

Cassell E (1995) Medicine, art of. In: Post SG (ed in chief) Encyclopedia of bioethics, 3rd edn. Macmillan Reference USA, Farmington Hills, pp 1732–1738

Cassell E (2004) The nature of suffering and the goals of medicine, 2nd edn. Oxford University Press, Oxford

Chang H (2004) Inventing temperature: measurement and scientific progress. Oxford University Press, New York

Curtis JR, Patrick DL, Shannon SE et al (2001) The family conference as a focus to improve communication about end-of-life care in the intensive care unit: opportunities for improvement. Crit Care Med 29:N26–N33

Daston L, Galison P (1992) The image of objectivity. Representations 40:81–128

Deber RB, Kraetschmer N, Irvine J (1996) What role do patients wish to play in treatment decision making? Arch Intern Med 156:1414–1420

Douglas H (2004) The irreducible complexity of objectivity. Synthese 138:453–473

Douglas H (2009) Science, policy, and the value-free ideal. University of Pittsburgh Press, Pittsburgh

Dupré J (1993) The disorder of things. Harvard University Press, Cambridge, MA

Ende J, Kazis L, Ash A et al (1989) Measuring patient's desire for autonomy: decision making and information-seeking preferences among medical patients. J Gen Intern Med 4:23–30

Engel G (1977) The need for a new medical model: a challenge for biomedicine. Science 196:129–136

Evans D (2006) Aristotle on the relation between art and science. In: Kuçuradi I, Voss S, Güzel C (eds) The proceedings of the twenty-first world congress of philosophy, vol 10, pp 21–30

Forstrom L (1977) The scientific autonomy of medicine. J Med Philos 2:8–19

Frank AW (1995) The wounded storyteller: body, illness, and ethics. The University of Chicago Press, Chicago

Gorovitz S, MacIntyre A (1976) Toward a theory of medical fallibility. J Med Philos 1:51–71

Hesslow G (1993) Do we need a concept of disease. Theor Med Bioeth 14:1–14

Hofmann B (2003) Medicine as *Techne* – a perspective from antiquity. J Med Philos 28:403–425

Jaeger W (1957) Aristotle's use of medicine as a model of method in his ethics. J Hell Stud 77:54–61

Kaufman SR (2005) …And a time to die: how American hospitals shape the end of life. Scribner, New York

King NMP, Churchill L, Cross AW (eds) (1988) The physician as captain of the ship: a critical appraisal. D. Reidel Publishing Company, Boston

Lacey H (1999) Is science value free? Values and scientific understanding. Routledge, New York

Lennox JG (1995) Health as an objective value. J Med Philos 20:499–511

Levinson W, Kao A, Kuby A, Thisted RA (2005) Not all patients want to participate in decision making – a national study of public preferences. J Gen Intern Med 20:531–535

Longino H (1990) Science as social knowledge: values and objectivity in scientific inquiry. Princeton University Press, Princeton

Lynch H, Lemon S, Durham C et al (1997) A descriptive study of BRCA1 testing and reactions to disclosure of test results. Cancer 79:2219–2228

Malterud K (1995) The legitimacy of clinical knowledge: towards a medical epistemology embracing the art of medicine. Theor Med Bioeth 16:183–198

Matloff E, Shappell H, Brierley K et al (2000) What would you do? Specialists' perspectives on cancer genetic testing, prophylactic surgery, and insurance discrimination. J Clin Oncol 18:2484–2492

Munson R (1981) Why medicine cannot be a science. J Med Philos 6:183–208

Nagel T (1979) Mortal questions. Cambridge University Press, Cambridge

National Cancer Institute (2011) NCI: http://www.cancer.gov. Accessed 28 Apr 2011

Nelson LH, Nelson J (eds) (1996) Feminism, science, and the philosophy of science. Kluwer Academic Publishers, Lancaster

Nordenfelt L (1987) On the nature of health. Reidel, Dordrecht

Parry R (2009) Episteme and Techne. In: Zalta E (ed) The Stanford encyclopedia of philosophy. Available at: http://plato.stanford.edu/archives/spr2009/entries/episteme-techne

Proctor R (1991) Value-free science? Purity and power in modern knowledge. Harvard University Press, Cambridge, MA

Ray J, Loescher L, Brewer M (2005) Risk-reduction surgery decisions in high-risk women seen for genetic counseling. J Genet Couns 14:473–484

Rubenstein W (2001) The genetics of breast cancer. In: Vogel V (ed) Management of patients at high risk for breast cancer. Blackwell Science, Malden, pp 19–55

Saunders J (2000) The practice of clinical medicine as an art and as a science. Med Humanit 26:18–22

Schaffner KF (1993) Discovery and explanation in biology and medicine. University of Chicago Press, Chicago

Schaffner KF (1999) Coming home to Hume: a sociobiological foundation for a concept of "health" and morality. J Med Philos 24:365–375

Schenker Y, Tiver GA, Hong SY et al (2012) Association between physicians beliefs and the option of comfort care for critically ill patients. Intensive Care Med 38:1607–1615

Stiggelbout AM, Kiebert GM (1997) A role for the sick role – patient preferences regarding information and participation in clinical decision-making. Can Med Assoc J 157:383–389

Teno JM, Clarridge BR, Casey V et al (2004) Family perspectives on end-of-life care at the last place of care. JAMA 29:89–93

Turner N, Tutt A, Ashworth A (2004) Hallmarks of "BRCAness" in sporadic cancers. Nat Rev Cancer 4:1–6

Waymack MH (2009) Yearning for certainty and the critique of medicine as "science". Theor Med Bioeth 30:215–229

Weinstein MC, Fineberg HV (1980) Clinical decision analysis. W.B. Saunders Company, Philadelphia

White DB, Malvar G, Karr J et al (2010) Expanding the paradigm of the physician's role in surrogate decision-making: an empirically derived framework. Crit Care Med 38:743–750

The Naturalization of the Concept of Disease

Maël Lemoine

Abstract Science starts by using terms such as 'temperature' or 'fish' or 'gene' to preliminarily delimitate the extension of a phenomenon, and concludes by giving most of them a technical meaning based on an explanatory model. This transformation of the meaning of the term is an essential part of its *naturalization*. Debating on the definition of 'disease', what most philosophers of medicine have examined is the pre-naturalized meaning of the term: for that reason they have focused on the task of delimiting disease and non-disease (health), mainly used conceptual analysis as a method of choice, and considered the nosological level of 'disease judgments' rather than the pathophysiological or psychopathological level of disease mechanisms, thus making them impervious to most scientific discoveries. By focusing instead on the naturalized concept of disease and following some suggestions by philosophers of biology and scientists in cutting-edge fields of biomedical research, they could garner results from a comparison of the mechanisms of diseases. This would ultimately result in a general theory of disease linked with our most general theories on living beings, among them, systems biology and network medicine. Before undertaking such a task, preliminary questions arise: is it likely that there are biological features common to different types of disease? Is it a philosopher's job to determine what they consist in? What use would such a general theoretical definition of disease be?

Introduction

A definition of 'disease' is often considered the foremost question in philosophy of medicine. Most philosophers of medicine who have attempted to capture the meaning of 'disease' in medical science (Boorse 1977; Wakefield 1992) have tried to draw it from so-called "disease judgments" (Boorse 1997) rather than from what

M. Lemoine (✉)
University of Tours, INSERM U930, IHPST, Paris, France
e-mail: lemoine@univ-tours.fr

P. Huneman et al. (eds.), *Classification, Disease and Evidence*, History,
Philosophy and Theory of the Life Sciences 7, DOI 10.1007/978-94-017-8887-8_2,
© Springer Science+Business Media Dordrecht 2015

could be called "disease explanations". A "disease judgment" is a judgment that some condition *is* or *is not* a disease, ideally grounded on some explicit, uncontroversial nosography. A disease explanation, on the other hand, could be defined a causal model of what is happening in a specific pathological condition, and it belongs to pathophysiology. To date, philosophers have considered their main question to concern the *concept* of disease, which would fulfill the role of a primary criterion of what should count as a disease, belong to nosography and, possibly but not necessarily, have a clear pathological explanation. I suggest that this is a misleading methodology. If one is to search for a definition of 'disease' *in science*, one should consider what concept, if any, is at work in disease *explanations*, for that is the place where medicine rightfully claims to be a science.

The question is therefore not, "in what sense do we call all those conditions the same, i.e., 'diseases'", but rather, "are there natural properties that mechanisms, or the series of events behind most so-called 'diseases', have in common?" Instead of drawing conceptual or trivial features from armchair philosophy, such as: "they involve suffering" or "they involve the failure of a statistically normal part-function", the quest is for natural properties drawn from the lab, or at least from scientific theoretical hypotheses, that hopefully sketch what a non-trivial definition of disease would look like, if such sketch is possible.

Before fleshing out this basic proposal, I ought to point out that I consider it to be compelling in its first two, mostly critical parts, but tentative in its third, constructive part. 'Tentative' here means that *in medicine*, it relies on hypothesis or even speculation, and that *in philosophy*, it faces many objections that largely explain why philosophers have sought for a definition in judgments rather than in explanatory theories. I will try to answer them in the course of this chapter.

My criticism of the traditional approaches in philosophy of medicine is based on a single distinction between pre-naturalized terms, i.e., terms that only refer to phenomena, and naturalized terms, i.e., terms that also explain how these phenomena occur. I consider the current debate on the definition of disease to be focusing on a notion of 'disease' as a pre-naturalized term, without considering the possibility that it might be naturalized. In the first part of this chapter, I introduce this distinction, and in the second part, I show how it enlightens the debates on the definition of 'disease'. Turning to the alternate proposal, I try to garner some results from science and from philosophy of biology in order to sketch out a search for a naturalized definition of 'disease'.

The Naturalization of Scientific Language

I begin with a distinction between 'pre-naturalized' and 'naturalized' terms, and the consequent definition of the process of 'naturalization'. Along with 'naturalism', these terms have been endowed with various meanings in various areas of philosophy. In the philosophy of medicine, 'naturalism' is widely used, not 'naturalization'. I stand as a naturalist, siding with Boorse and others in giving priority to a conception of disease as a natural fact rather than a normative one. Yet my argument comes up against a method used by naturalists and normativists alike: conceptual analysis of 'disease judgments'. For that reason I must clarify in what sense the position I am advocating is one of naturalism.

It is naturalism in a sense very similar to Quine's; an influential standpoint in the philosophy of science at large, yet strangely ignored in the philosophy of medicine. Its basic tenet is that

> it is within science itself, and not in some prior philosophy, that reality is to be identified and described. (Quine 1981, 21)

There is nothing here that naturalists in the philosophy of medicine would not accept, given that to them, "some prior philosophy" would then mean exactly the sort of things normativists are engaged in. My point is that naturalists too are committed to "some prior philosophy" in supposing, if implicitly, that the way reality is to be *identified* mostly does not change according to what is being *described*. In that, they seem to admit of a genuinely independent philosophical stance that they, apparently modestly, do not conceive of as prescribing standards to science, but claim is nothing but descriptive. The kind of naturalism I am defending here does not admit any independent philosophical stance on the notion of disease. What we, as philosophers, should be trying to do is exactly what scientists would try to do, were they to consider the search for a general definition of disease worthy of their time. Gathering their results on specific diseases and types of disease, and reformulating them into the simplified form of a theory, is exactly what naturalistic philosophy of medicine should attempt to achieve. This is not 'describing' science, which supposes some external point of view, but trying to do some science, with very low likelihood of any useful outcome.

Many, including Boorse himself, have considered that the term 'naturalism' in the philosophy of medicine is rather unfortunate, and should be replaced by 'descriptivism' (Boorse 1997, 102). The problem is that in a sense, normativists too are descriptivists. What I suggest then is to distinguish a naturalist notion of disease from a *naturalized* notion of disease. In other words, in naming phenomena 'diseases', we surely have some preliminary, if vague, notion of what we are referring to (the notion of disease in general). But in collecting facts about these phenomena, we end up with a resulting, albeit evolving, knowledge of how these phenomena work, and possibly, with a revised notion of what we had called 'disease' in the first place. Naturalization implies focusing on the resulting notion rather than on the preliminary notion. Focusing on the preliminary notion of things is exactly what earlier contributions to the philosophy of biology once did, elaborating what laymen and biologists alike first think of when they hear terms such as 'life', 'organism', or 'function'. Philosophy of medicine appears to do just the same. A naturalist in the philosophy of medicine trying to define 'disease' should not gather terms from a common understanding of disease and try to refine the meaning – defining what 'function', 'organism', 'species', or 'normal' mean for instance. Instead, she should investigate biomedical research on diseases and look for more general facts and theories that possibly define disease in general.

Pre-naturalized and Naturalized Terms in Science

There are probably many properties of naturalized terms in science: mathematization, reference to natural kinds,[1] etc. What is needed here to make the important point is only one of them: the fact that they refer to an explanatory model, as opposed to pre-naturalized terms referring to observed phenomena as such.

As a matter of fact, some terms are accepted into the scientific language because they refer to a class of phenomena considered to be of interest to a particular scientific discipline. In present-day chemistry for instance, 'gold', 'copper', 'iron', and 'oxygen' are such terms, not 'orichalcum' or 'phlogiston'. 'Magnetism' now refers only to a subclass of the alleged phenomena it used to refer to in the eighteenth century, excluding what was then called 'animal magnetism'. 'Transmutation' has today an altogether different sense than it had in alchemy. To denote a class of presumably interesting phenomena is one function of all these terms.

Another function of a scientific term is to link the understanding of this class of phenomena with reference to a theory or to an explanatory model. For instance, whenever the term 'homology' is used in evolutionary biology, the phenomena thus referred to are considered to be the result of common lineage in a Darwinian framework. An explanation is supposed – evolution by natural selection – and one must know this explanation in order to understand the term correctly. The term that fulfills that function belongs to the "theoretical vocabulary" (Hempel 1977; Suppe 1977) of a science. The actual word is not necessarily forged anew for that purpose: the same word can fulfill both functions.

However, the meaning of a term endowed with the first function only, i.e., to refer to a class of phenomena, changes when it becomes endowed with the second function too, i.e., to refer to an explanatory model. For instance, 'temperature' first referred to a class of observable properties (being more or less cold, warm or hot). The class may not change, but the meaning certainly does, when 'temperature' is defined theoretically in the kinetic theory of gas and in thermodynamics, at least in that a new, more relevant meaning is attributed to the term.

As a matter of fact, when a scientific, technical definition is given to a term such as 'temperature', the class of phenomena thus referred to sometimes changes. Sometimes it becomes an empty set, as in the case of 'phlogiston' (Enç 1976). Sometimes its unity is questioned: 'fish' is now a paraphyletic group of various taxa, and its extension does not correspond anymore to everything that lives and perhaps swims in the sea (including whales and dolphins for instance) – cladists even consider that 'fish' does not refer to any real group. Sometimes the extension of the term is just modified: in that case, the new sense of the term being considered the true one, we generally refer to older uses (i.e. before the theory became available)

---

[1] Although the examples given here are very similar to those given in all the discussions about natural kinds, it would require a great deal of discussion to use this very polysemic phrase here. While it is obviously possible to define 'natural kinds' in a way that serves the purpose of this chapter, I doubt it is the most straightforward way to achieve its goal.

as gross forms of the newer use, if not erroneous forms (Putnam 1975, 310–2). Sometimes, no extensional change follows, as in the example of 'temperature'.

Naturalization

Given the preceding distinction, the phenomenon of naturalization comes into focus. Again, there are many properties attached to it. For instance, in the philosophy of medicine, we have an intuitive grasp of what it means in phrases such as 'the naturalization of mental disorders', i.e. their transformation into biological entities, and 'the naturalization of diagnostic reasoning', i.e. its transformation into a natural object to be investigated by cognitive science. Hippocrates naturalized epilepsy a long time ago by defining it as an ailment of the brain, not as possession by some god. Infectious diseases were naturalized with, or maybe before, the emergence of the germ theory, and so on. There are probably many aspects of naturalization in medicine, but we are interested here in only one of them. In naturalization, a referential term is transformed into an explanatory term with reference to a causal model or a theory. It is not necessary, but it is possible, that a naturalized term comes from common language first, and that its extension changes in the process. However, with the function of the term, its meaning invariably changes. Note that this model refers not only to causes, but to natural causes – as opposed, for instance, to the will of the gods or to the hopes and fears of human-kind. Note also that naturalization does not require knowing the right cause, whatever this may mean, but only to refer the term to natural causes that explain the phenomenon it designates.

Pre-naturalized terms are not exactly pre-scientific terms: they rather belong to what Hempel calls the "pretheoretical vocabulary (…) relative to the theory in question" (Hempel 1977), oftentimes, terms from previous theories. Further, their function is strategic: they reflect a choice by scientists to consider a class of objects to be natural, important or fruitful, before even knowing whether the generalizations based on this class of objects can support further theorization. Ultimately, the definition of pre-naturalized terms is bound to be imprecise. As a matter of fact, what matters most is the fact that we still call '*x*' what we always have, not that '*x*' is given sharp criteria of delimitation: hence the resort to complicated sets of necessary and sufficient conditions when we try to define *things*. Achinstein described the general features of such definitions under the label of definition "of physical objects or stuff" (Achinstein 1968, 2).

Naturalized terms in turn are entrenched in explanatory models. They still refer (more or less) to the same class of object, no longer grouped by their manifest features, but rather by their structure or inner working (Thompson 2011). Besides their explanatory function, naturalized terms still delimitate the class of objects. Nevertheless, the kind of definition is not the same anymore: theoretical criteria are precisely determined through necessary and sufficient conditions, for what they focus on is not the natural class of observed objects anymore, but the model these objects are supposed to be similar too (Achinstein 1968, 2).

Lingering on a Pre-naturalized Definition

Again, naturalization consists of manifold processes, among which only one has been defined here. This is because this one suffices to make my main critical point against the traditional debates over the definition of disease. This point is that the traditional debate is only concerned with reference, not explanation, i.e. it lingers on a pre-naturalized notion of disease in general instead of looking for a naturalized one (section "The debate over the definition of 'disease' is about a pre-naturalized term"). What is at stake is not the naturalization of various diseases, but the naturalization of the very notion of disease, if this is possible. At first sight, it seems that the naturalism-normativism debate in the philosophy of medicine is about whether 'disease' is a naturalized or a pre-naturalized term. As a matter of fact, this is not really the case: on the contrary, it is about the pre-naturalized term 'disease' (section "The opposition of 'naturalism' and 'normativism' about disease is about a pre-naturalized term"). This leads to a paradoxical standpoint for so-called 'naturalism' in the philosophy of medicine, mainly, its ignorance of contemporary knowledge about diseases in particular and also possibly, disease in general, assuming somehow that the latter is perennial despite scientific progress (section "The consequences for naturalism of focusing on the definition of a pre-naturalized term").

The Debate Over the Definition of 'Disease' Is About a Pre-naturalized Term

In the philosophy of medicine, one of the most heated debates has been around the definitions of 'health' and 'disease'. I will consider here 'disease' only, so as to simplify the matter. Many philosophers have proposed and discussed definitions (Boorse 2011), while others have discussed either the possibility (Sadegh-Zadeh 2000) or the usefulness of such a definition (Hesslow 1993). In this section, I present my reasons for thinking that these debates were focused on pre-naturalized terms, and the consequences of this. From the view I propose, this focus follows from two widely shared premises in the field:

- A conceptual examination of 'disease' by a philosopher should limit itself to the meaning of the term as it is;
- There is no general theory of disease in medicine, and therefore no naturalized concept of disease (at least, not yet).

To put it differently, conceptual analysis cannot be achieved on a naturalized concept of disease that does not yet or will never exist. On the other hand, there is a pre-naturalized concept of disease, which mostly fulfills the function of determining what is pathological and what is not. It is only this that it is a philosopher's task to describe.[2]

---

[2] Nordenfelt distinguishes a *conceptual definition* and an *empirical theory* of disease (Nordenfelt 1995, 9). According to him, whereas a definition characterizes a concept, an empirical theory characterizes "the phenomenon represented by the concept". From an analytic point of view,

I take the second premise to be obvious, at least in the sense that in medicine, there is no science of disease *in general*. As previously stated, specific diseases such as diabetes, epilepsy, cancer, and asthma, have been naturalized successfully: their inner workings are accounted for, fully or partially, by causal, mechanistic models explaining (some of) their various presentations. But there is no corresponding general model or theory accounting for disease in general, and maybe none is possible (or even useful) because of the diversity of specific diseases.

It is the first premise that I question here. Such a descriptive stance on the meaning of 'disease' as it is, generally implying the kind of examination that is called 'conceptual analysis', I have already analyzed at length in two papers (Lemoine 2013, forthcoming). As an important result of this premise, the criterion of a proper feature for defining 'disease' is its presence in everything (consensually) called a disease, and absence in consensual non-diseases. For instance, 'pain' (as such) is not a good feature both because there are many painless pathological conditions such as being paralyzed and because there are painful non-pathological conditions, such as giving birth or teething. Plain statistical abnormality is not a proper defining feature for disease either, both because diseases can be widespread and non-diseases rare, etc.

Taken as such, this is a very narrow interpretation of what could reasonably be called 'conceptual analysis' in philosophy, one that seems to equate it with 'meaning analysis' (employing a very specific and debatable method) and to eliminate, for instance, the explication of the conceptual relations between terms or the tentative reformulation of vague notions into more precise concepts. Yet it is obviously what philosophers of medicine call 'conceptual analysis' (Lemoine 2013), and it is rather consistent with what Hempel said about it (Hempel 1952). It is also very similar to what opposed Millikan and Neander in the philosophy of biology about the definition of 'function' (Millikan 1989; Neander 1991).

Thus, the focus of the traditional debate is obviously on the first function of a definition in science, that of referring to, and delimiting, a class of objects. Two associated indications of that are: first, that conceptual analysis seems the most natural method for delimiting diseases from non-diseases, a goal many philosophers of medicine would consider to be theirs; second, that the naturalized knowledge about these particular diseases is generally not relevant to answer the question why they are diseases as these philosophers understand it. For instance, Wakefield questions whether some states labeled 'major depressive episodes' by the *Diagnostic and Statistical Manual of Mental Disorders*, are really pathological (Horwitz and Wakefield 2007). Yet almost nothing known or hypothesized about the workings of depression, mental or biological, is thoroughly examined for itself, but the general idea that a pathological condition is the absence of a trait that has been selected for, and for the criterion of the clinically assessed relevance of the reaction to the context. The same goes for Boorse's analyses of various diseases. It apparently

---

philosophers engage (or should engage) in definitions, not theories. To some extent, the distinction implies certain independence between both accounts, so that at least two persons may very well have distinct empirical theories about the phenomena of health and disease, and still share the same concept.

suffices to know that some sub-function somewhere, the ins and outs of which do not matter, is impaired and thereby fosters the organism's ability either to survive or to reproduce, and so much for the pathophysiological details. The kind of analysis both Wakefield and Boorse are engaged in does not require much knowledge of physiological, psychological or pathological mechanisms, notwithstanding their declared naturalism (and their own knowledge of these questions). As Murphy puts it,

> the approach (…) tries to give the meanings of terms without investigating what in the world those terms actually refer to. (Murphy 2006, 52)

The reason for this is that these philosophers consider nosology, i.e., a list of conditions that should be classified as diseases, not pathophysiology or psychopathology, the natural science of the inner workings of actual dysfunctions in the human body or mind, to be the test criterion of a good definition. They need that list to provide examples, and all they need to know about these conditions is whether their proposed definition can or cannot be applied to them. A first problem with this position is that it provides little help in considering major issues that nosology actually encounters, which is not whether a particular condition is a disease, but rather how diseases form classes or types such as: infectious, genetic, mental, neurological, and so on, and according to which criteria – etiological, anatomical, and so on.[3] The second and most important problem here is that the features these philosophers propose to define 'disease' are very general, and so is the level of knowledge on particular diseases required. This sounds paradoxical for naturalists. As a matter of fact, if they ventured to advance more specific features of diseases, say for instance, that *in every one of them* there is a determined enzymatic phenomenon defined as…, then there would be much more to know about particular diseases to test this hypothetic definition.

The Opposition of 'Naturalism' and 'Normativism' About Disease Is About a Pre-naturalized Term

'Conceptual analysis', or 'meaning analysis', is often understood as capturing 'what people have in mind' when using a given word. What I previously called the term's function remains implicit from this point of view. The main opposition in the debate on disease, normativism versus naturalism, is about the meaning of the term as a means to delimit the class of diseases. According to normativists, the conditions referred to in this class are heterogeneous from a biological point of view, and uniform only from the point of view of their specific consequences for human existence (harm, disability, and so on). To naturalists on the other hand, the class is homogeneous when considered as a group of subtypes of biological phenomena implying dysfunction.

The opposition of naturalism and normativism is not the opposition between defining a naturalized term and defining a pre-naturalized term or concept of disease.

---

[3] Thanks to Marie Darrason for drawing my attention to this point.

Both positions are in fact focused on the pre-naturalized concept of disease. The bone of contention is about the kind of features that actually describe the class of diseases: do the correct features sound scientific, biological, physiological, or do they refer to human condition, praxis, hindrances to aspirations and achievements? This could be phrased more precisely in the following manner: are the defining features of 'disease' *likely to be naturalized*, as expressed by terms such as 'dysfunction', 'adaptation', 'fitness', 'evolution', 'statistical', and so on, or are they not, as expressed by terms like 'inability', 'evil', 'happiness', 'harm', and so on?

As a matter of fact, as long as there is no general theory of disease, it is possible to think that there can be no consistent grouping of diseases after naturalizable features, no concept of disease as a natural fact, no single natural definition of disease. Indeed, normativism seems to require this state of affairs. For, if the concept of disease was naturalized, it would become hard for a normativist to make a case. One could still gather imaginary cases (in the 'thought experiment' style) where we would readily think that something not fulfilling the naturalized conditions for being a disease is a disease, or the contrary. One could still assume that the concept of disease was initially formed as a normative means for classification. Eventually, one could melt one's position down into a general constructivist stance, claiming that we were looking precisely for a class of phenomena we disvalued in the first place, and so on. Nevertheless, as there would be this robust theory of a whole class of phenomena, the scientific notion of disease would now stand on its own feet. The situation would be the same as it has been for the acceptance of the consequences of evolutionary theory: some, despite accepting the evidence that humans and other animals share a common ascent dating from some remote time in the past, would still assume that this cannot be what we understand when we are talking about 'ascent', and that somehow, human ascent is very special. They would make their case, for instance, by positing beings of another ascent that we would still call humans, etc. It is conceptually defensible, but what is the point? 'Ascent' is nevertheless a conceptually consistent scientific notion. If 'disease' was thus successfully naturalized, it would similarly be a non-starter to claim that there is no consistent scientific concept of disease. Yet, of course, this remains an open question.

The Consequences for Naturalism of Focusing on the Definition of a Pre-naturalized Term

It is strange enough for naturalists not to bother about more details of what is known of the processes at work in various diseases than their own proposed definition requires. What is more embarrassing still is the consequence that major discoveries on how diseases work would not have any effect on their proposed definition, as long as the class of diseases remained the same. The discourse on 'function' and 'dysfunction' seems to be so general that no empirical discovery could impinge on it, and that on the contrary every new result is somehow bound to fit with that very general conceptual framework. Whatever is discovered, say for instance that in microbiology it was shown that *every* disease actually implies a microbial signature specific to diseases (e.g. unbalance in the various microbial population inhabiting the human body), which would provide a powerful naturalized definition of disease, we could probably still think of disease as a part-dysfunction.[4] To be sure, this captures one traditional way of thinking about disease in medicine, but the fact that it is not at odds with other ways of thinking about disease does not mean that it is all that physicians have in mind when they use the term.

It would probably be exaggerated to worry about the power of conceptual analysis to hinder scientific progress by imposing some kind of conceptual straightjacket on the term 'disease'. At least in Boorse's case, possibly in Wakefield's too, one cannot assume that this conceptual straightjacket is the commonsense notion of disease:

> Wakefield is analyzing a concept that plays a role in commonsense thought and arguing that the task of science is to identify the natural processes that accord with that commonsense concept. We argue that this represents an attempt to use conceptual analysis to legislate what should be acceptable science. (Murphy and Woolfolk 2000, 271)

Although they might resist the introduction of new conditions or their exclusion in the name of their definitions, all naturalists do not claim to impose on further developments of the notion of disease itself. For the time being, Boorse says, the definition is adequate, and when it will not be anymore, it will being changed (Boorse 1997, 53). How are we to know that a previously correct analysis is not anymore, and that the problem is not with new conditions hitherto wrongly considered to be diseases? Boorse's answer would probably be that scientists will decide and that philosophers should follow. What would philosophers do if some naturalized definition such as that of an unbalance of microbial populations in the human body came to redefine disease, and was at odds with the received, conceptual analytic view on disease? Most probably, Boorse would admit in the same way that his conceptual analysis is not correct anymore. Wakefield's position would probably be more delicate, for he seems to conceive of conceptual analysis as a way to resist abusive medicalization.

Now, what would conceptual analysts do if a naturalized definition of disease emerged which would *not* be at odds with their conceptual definition? Most probably they would stick to the latter and admit the former. This is precisely what is wrong with naturalism as it is: any possible definition goes, provided that it is successful in conceptual analysis and phrased in terms of natural science, however vague. This is the case because naturalists focus on the prenaturalized, existing notion of disease – 'disease' as a term with no reference to any general explanatory model. For this reason, conceptual analysis can be a conceptual straightjacket, but my concern is rather that it is a loose shirt.

---

[4]The notion of disease is currently considered to be multiply realizable. In the microbiome *Gedankenexperiment*, it would still be. Yet instead of being just a concept, disease would refer to a specific fact in nature. The multiple realizability of one specific disease is another, similar question.

A Sketch of What a Naturalized Definition Would Look Like

The search for a naturalized concept of disease is obviously more sympathetic to naturalism than to normativism, but it does not share its reliance on conceptual analysis. I advocate the search for a naturalized, theoretical definition of disease rather than the pursuit of the conceptual analysis of the term. Much of what can be offered to that end here is just lifting some objections that stand in the way of such an enterprise. This begins with questioning the idea that a definition of disease in general can result from the analysis of linguistic use only, not from a general theory of disease (section "Questioning the implicit science-philosophy division of domains"). As a matter of fact, a naturalistic stance is not necessarily restrained within the boundaries of conceptual analysis, but can extend to the theoretical use of natural features of disease as proposed by science, an approach I call 'inductive', of which I provide a short example (section "Inductive approach to the empirical properties of disease"). Such an inductive approach necessarily leads to a second question: whether explanatory models of dysfunction endow them with specific features never to be encountered in models of normal functioning (section "Are there features specific to disease in explanatory models?"). A third question is that of the link between two general properties: being a living organism, and being subject to diseases. Here again, I would advocate a naturalized approach to what 'living organism' means, but not reject theoretical consideration such as what theoretical biology has to say about a general theory of disease (section "Is disease definable at the level of theoretical biology?"). Finally, I briefly consider a possible shift in the notion of a function and its possible consequences for that of a dysfunction: can it renew the traditional theoretical template of medical science (section "A new disease template for disease sciences?")?

Questioning the Implicit Science-Philosophy Division of Domains

Usual definitions of 'disease' come in three different forms, as definitions of disease x (i.e. diabetes, cancer, flu, …), as definitions of types of disease (i.e. infectious, metabolic, mental, … disease) or as definitions of disease in general. I take for a fact that there is as yet no general scientific theory of disease. The consequence is that no unified naturalized concept of what disease in general consists in is currently available in medical science. This is good enough reason for philosophers to have focused on a pre-naturalized concept of disease instead, provided that all there is to philosophy of science is conceptual or meaning analysis.

Another, related claim is that scientists actually display no interest whatsoever in a general theory of disease, and that they focus instead on particular theories of diseases (Thagard 2006). This correctly reflects what is considered to be known for facts so far in medicine, but omits the many hypotheses research generates about diseases.

From these two claims, one can draw the following received view:

- Definitions of *particular* diseases and types of diseases rely on theories or models and belong indeed to scientific thinking exclusively;
- Definition of disease in *general* relies on conceptual analysis and belongs to philosophical thinking exclusively.

I can see three main reasons why this divide is questionable. First, philosophers could try conceptual analysis on specific diseases or types of disease as they do with 'mental disorders'. Most philosophers probably deem a conceptual analysis of particular disorders such as diabetes pointless. Some might not agree though (Engelhardt Jr. 1975; Severinsen 2001; Horwitz and Wakefield 2007). The same goes for the concept of disease in general: most consider that a concept of disease must exist if we are not to gather particular diseases randomly under the same label, and some defend the view that it relies on a *petitio principii* (Sadegh-Zadeh 2000).

Second, I suggest that there is no a priori reason to think that it is impossible to apply both methods, i.e. naturalization and conceptual analysis, to the definition of either particular diseases, types of disease or disease in general. It can be the case that a formulation of such a definition does not rest upon lexical analysis alone, but is based on a general theory of disease as a particular natural phenomenon.

Third, the fact that such a theory does not exist yet is not an obstacle if philosophers do not consider their task to be *describing* science, but rather *contributing* to science, by formulating as tentative theories whatever generalization the science of diseases produces. Description is conservative and there is no point in trying to do it in a constantly changing domain. This is why philosophers of science tend to rely upon parts of medicine that do not seem to change much, typically, semiology instead of pathophysiology. Let us consider, for instance, that a philosopher of science is engaged in defining a particular disease, type 1 diabetes. If she tried conceptual analysis, the main features she would rely on are clinical, because those do not readily change. Research, on the other hand, shows the entrenchment of mechanisms diabetes consists in: it can be considered either an auto-immune disease, or a metabolic disease, or a genetic disease, and who knows what else. Another example is the investigation of the links between vascular diseases, Alzheimer's disease and late onset depression. Although there are interesting hypotheses about the nature of these links, a traditional approach still prevails in therapeutics, where they are respectively considered, say, as a dysfunction of the arterial system, a degenerating process and a problem with monoamine levels. A philosopher looking for a naturalized definition of a particular disease would try to encompass all we know about the inner mechanisms of this disease in order to define it, rather than analyze orthodox usage of the term as it is defined in semiology. This is very likely to be premature and to result in failure, but it is also very likely to contribute to science in a way scientists do not often contribute themselves. It is all the more true for the search of a type of disease or disease in general. As a matter of fact, research, not its application, should be where philosophers of medicine start if they conceive of themselves as philosophers of science.

Inductive Approach to the Empirical Properties of Disease

The latter result implies that naturalists using conceptual analysis of a pre-naturalized term have worked *deductively*. Instead of looking for what medical science has actually discovered about disease, they have more or less checked whether their assumptions hold or looked for features general or vague enough to hold whatever science discovers in the near future, such as: 'dysfunction', or 'impairment to normal functioning as designed by evolution'. On the contrary, working on the naturalization of disease implies to work *inductively*, that is, take into account as many details as are available about diseases, select the right ones, and look for what science itself suggests might be unified properties of disease in general. These properties are not trivially conceptual, but they are theoretical in the sense of scientific theories of nature. By trivial or conceptual properties, I mean such predicates as 'survives' or 'reproduces', as defining features of living beings in general. Theoretical properties are either empirical properties such as 'contains DNA or RNA', or explanatory properties such as 'evolves' in a Darwinian sense, in the case of a living being.

Yet considering the whole field of alleged pathological conditions, "a congenital clubfoot, a sexual inversion, a diabetic, a schizophrenic", as Canguilhem once put it (Canguilhem 1991), one hardly deems it possible to gather empirical features common to all. One part of the solution is: not trying to subject natural facts to lexical constraints. When science discovers interesting common mechanisms in various phenomena, it sometimes contradicts the previous use of a term to denote them. The same goes for a naturalized theory of disease: were it not to include congenital clubfoot and hypertension or schizophrenia, but diabetes, cancer, neurological, infectious, metabolic and auto-immune diseases, plus a few other categories, including what is now considered 'normal aging' for instance, it could stand for a general theory of disease anyway.

Another objection is that there have been many attempts at general, non-trivial natural theories of disease: Hippocrates' theory of the unbalance of humors, the iatromechanists' theory of mechanic disruption, Broussais' theory of irritation, Groddeck's psychosomatic theory. All were failures. On the other hand, the view that medicine started progressing precisely when physicians stopped searching for a general theory accounting for all diseases and focused instead on the explanation of each particular disease has a very strong rationale. As a matter of fact, what made these attempts failures is probably not generality, but prematurity. How would we know before trying whether it is the case nowadays? Moreover, the search for generalization is not incompatible with the search for specific explanations: although it apparently makes it more difficult because of the great diversity of what is known about particular diseases, it alone also provides the means to make correct or powerful generalizations. This is the case for instance each time a new mechanism is discovered in one disease and unexpectedly also happens to be of importance to other diseases.

An example of such a descriptive, inductive work on empirical properties of disease would be to incorporate the contribution of network medicine to the naturalization of disease.[5] *Network medicine* consists in the study of biological networks implied in diseases (Barabási et al. 2011). One kind of biological network is the interactome: the gigantic interaction of protein-coding genes, RNA, proteins and metabolites, of which sub-networks are gene networks, protein networks, and so on. Within the interactome, 'modules' consist in clusters of components with relatively more interactions between them. Disease phenotypes are associated with some of these modules, called 'disease modules'. Another kind of biological network in network medicine is the diseasome, that is, the graph of the links between all genetic diseases and disease genes (Goh et al. 2007).

What is to be said from the perspective of a naturalization of the concept of disease? First, this approach takes a definite step away from the regionalization of explanatory models of distinct diseases ('short-range definitions'), towards some form of generalization, in that it tries to link many diseases previously thought not to have anything causal in common, but only to have the conceptual property of being 'genetic diseases' in the broadest sense. Jimenez-Sanchez et al. consider that

> the functional classification of disease genes and their products will reveal general principles of human disease.(Jimenez-Sanchez et al. 2001)

Darrason also suggests that network medicine provides a "genetic theory of Disease" (Darrason 2013). According to her, it is a theory of *Disease* and not a theory of *diseases* insofar as it should be instantiated in each particular disease whatever its type, i.e., it does not necessarily reflect the classification of diseases currently received, but possibly classifies diseases according to various kinds of genetic mechanisms. It is a *genetic* theory of Disease in that it geneticizes diseases without genocentrism, i.e., without claiming that genetic mechanisms are the main causal processes involved in the explanation of diseases. Thereby it includes all diseases with a genetic component, that is, probably nearly all pathological conditions, injuries or intoxications excluded. In short, it contains a theory of the role of genes in the global biological phenomenon called 'disease'.

Second, network medicine exemplifies what kind of naturalized property of diseases science can discover. For instance, it draws on the distinction of 'essential genes', i.e. genes whose homologous recombination leads to death in utero, and 'non-essential genes' (Jeong et al. 2001). There are disease genes in both groups, in the proportion of approximately 1 'essential' for 3 'non-essential'. These non-essential genes in turn are more likely not to encode for 'hub' proteins, i.e., highly combinable and therefore ubiquitous proteins. In the end, such non-essential, non-hub proteins are much more likely not to be housekeeping proteins, that is, to be peripheral and not central from a functional point of view. This result, although highly speculative, is also theoretically important: it tends to prove that genetic diseases affect more peripheral functions than central functions in the self-maintenance machinery of the living being, and to quantify how much more. This is

---

[5]Another example would be microbiome medicine: the importance of the microbiome in the human organism is very likely to change our conceptions of at least some diseases, and possibly disease in general (Dupré 2011).

a naturalized property of genetic diseases as a whole, although it is not specific. The same kind of approach has since been applied to pleiotropic human disease genes (Chavali et al. 2010) and comorbidity (Park et al. 2009).

Third, network medicine shows how disease entities could be grouped in entirely new types. For instance, Jimenez-Sanchez shows that four functional categories of gene products, namely, transcription factors, enzymes, receptors and modifiers of protein functions, are significantly more involved in four kinds of diseases with different typical onset: in utero, 1 year old, till puberty, early adulthood (Jimenez-Sanchez et al. 2001). A more general framework for such a reorganization of nosology according to the results of '–omics' has been sketched in various papers (Loscalzo et al. 2007; Loscalzo and Barabasi 2011). From a philosophical perspective, Darrason examines the case of the classification of infectious diseases in the same perspective (Darrason 2013).[6] A renewed classification of types of diseases is likely to provide insights into a potential deep unity of disease, as it would be based on a single principle rather than on several heterogeneous principles.

Such properties of disease are not trivial, but empirical: they do not obtain by explicating what we usually mean by 'disease', but rather by observing what the properties of the phenomena themselves are. A philosopher does not contribute either as a direct observer of nature, or as an analyst of linguistic use, but as someone emphasizing and combining interestingly universal properties that could contribute to a general theory of disease.

Are There Features Specific to Disease in Explanatory Models?

After having considered whether some empirical properties of various diseases may provide a basis for a general theory of disease, I now turn to the question: are there theoretical features of disease such as a theoretical definition of 'dysfunction' that could be relevant to the task of naturalizing this concept?

In biomedical science, normal and pathological mechanisms respectively pertain to related, but distinct sciences. For the knowledge of normal, psychological/physiological facts does not by itself contain the knowledge of abnormal, pathological facts. In the philosophical debate over biological functions, it has often been overlooked that most potential ways in which a system might conceivably fail to do what it is supposed to are never observed nor can even be the case. For instance, there are many different structures for functional hemoglobin, many different structures for dysfunctional hemoglobin, but many more conceivable structures of dysfunctional hemoglobin than could ever be observed. It follows from that as well as from other arguments that the science of biological dysfunctions is not just

---

[6]Many more contributions could illustrate what naturalized predicates of a definition of disease could look like. Among them are ecological views on the involvement of the microbiome in health as in many diseases, sometimes unexpected (as is the case for the so-called "mind-body-microbial continuum"), but also Darwinian medicine (Nesse 2001) and its critiques (Germain 2012).

deducible from the science of biological functions. Although this is not the place to delve deeper into that question, overlooking that discards most of what has been said about functions in the philosophy of biology, as non-applicable to pathophysiology. Dysfunction is not just a reflection of function but a different kind of phenomenon, a fact of its own. I have discussed elsewhere what the opposition of the healthy and the pathological could mean logically (Lemoine 2009), and analyzed Canguilhem's, Boorse's and Nordenfelt's position on the question. In the end, trying to naturalize health could lead to an entirely different result than trying to naturalize disease.

A related question is whether the knowledge of the pathological nature of a fact comes from the knowledge of a set of normal facts that the former deviates from. It is true that we classify a condition as a disease in the first place because we disvalue it, not because we observe specific biological properties. Yet it is also true that we do not classify this condition as a disease only because it is harmful or unpleasant, but because we also suppose it implies dysfunction of some sort, before even knowing it. It is true again that in so doing we may wander for a long time – maybe there is nothing biologically dysfunctional in most cases of depression for instance. My question is: is there some notion of a pathological dysfunction that scientists have discovered or are discovering, after which it would be possible to check whether what we considered *in the first place* to be dysfunctional because it is harmful in some way, *really is* dysfunctional?

It would be the case if explanatory models of dysfunctional processes endowed them with specific properties never encountered in normal processes or explanatory models thereof. Focusing on mechanistic explanatory models, Nervi has proposed that pathological mechanisms are causal sequences of biological events with specific properties not displayed by physiological or normal mechanisms. They are not just a sequence of biological events strictly corresponding to the normal sequence except for one step at least, which is absent, impaired or altered. Whereas inductively discovered properties of disease are empirical, these are, or would be, system-related properties of disease.

According to Nervi, some diseases display *outcome variability*, i.e. the fact that

> their nature is not identified by specific outcomes; while a physiological mechanism has always a favourable outcome in normal conditions, a pathological one may result in outcomes ranging from death to complete recovery. (Nervi 2010)

They also display *no range constraint* of either internal or external conditions of working, whereas normal mechanisms necessitate relative invariability of background conditions. Lastly, they display *ambivalence*, that is, they possess

> the potential to work as adaptive or maladaptive in the presence of different regulating factors. (Nervi 2010, 221)

Nervi does not take these properties to be general, conceptual and defining properties of diseases as such, but rather some interesting natural properties of some pathological mechanisms that hitherto went unnoticed. Critiques (Gross 2011; Moghaddam-Taaheri 2011) have focused either on the existence of specific properties of explanatory models of diseases, or on the very properties Nervi proposed. These are distinct points.

Whatever the outcome of this debate, it is worth noting that the perspective is completely different than that proposed by conceptual analysis. For the discussion is about the kinds of features to be encountered in pathophysiological explanations, not just in symptoms or in gross ideas about how a human organism usually survives or reproduces. They also differ from the types of properties previously referred to and likely to be drawn from empirical science such as network medicine, genomics, microbiology, epidemiology and so on. Yet it is expected that both these empirical properties and these theoretical features, if compatible, should combine, but not necessarily result, in a complete theory of disease based on some general theory of organisms.

Is Disease Definable at the Level of Theoretical Biology?

There have been contributions to a theory of disease in theoretical biology, such as a general theory of organism. Only organisms suffer from diseases, and it seems that all organisms can suffer from diseases. Is it a universal fact, i.e., would it be naturally possible that some organisms cannot suffer from disease, or a theoretical necessity, i.e., any organism must be able to suffer from a disease because of what being an organism implies? In the latter case, theoretical biology provides essential ground for a theory of disease, whereas in the former case, it is more likely that diseases are just unrelated empirical facts with no overarching theory thereof.

Systems biology renews traditionally important notions for the definition of a disease, such as that of a function. Moreno, Mossio, and Saborido proposed replacing the concept of a function by an 'organizational' concept of function translated into the language of theoretical biology and complex systems theory (Mossio et al. 2009). Among systems, some are characterized by self-maintenance (Mossio and Moreno 2010), such as organisms, hurricanes, autocatalytic networks, and among these, some are also organizationally differentiated, that is, they produce

> different and localizable patterns or structures, each making a specific contribution to the conditions of existence of the whole organization. (Mossio et al. 2009, 826)

The baseline idea is that in a self-maintaining, organizationally differentiated system S,

> a trait T has a function if and only if:
>
> C1: T contributes to the maintenance of the organization O of S;
> C2: T is produced and maintained under some constraints exerted by O;
> C3: S is organizationally differentiated. (Mossio et al. 2009, 828)

In systems fulfilling C3, some traits fulfill C2 but not C1. They are *dysfunctional* traits and must be distinguished from *side effects* of traits that do not prevent these traits from fulfilling C1. On the one hand, the usual contribution of those dysfunctional traits is either not indispensable and can be compensated so that the system continues to self-maintain in the same way (that is, belong to the same class of self-maintaining systems), or indispensable, so that a new regime of self-maintenance is either required or impossible (which leads to destruction). Most non-fatal diseases belong to the second category: they involve a new regime of self-maintenance.[7] On the other hand, dysfunctional traits resort either to primary or to secondary functions, that is, contributions to either baseline or complex regimes of self-maintenance (such as respectively the heart's functions of 'pumping blood' and 'making diagnostically interesting noises').

Third, systems biology also provides us with some new potential features for a description of disease. Consider the notion of disease as a living process of its own, as opposed to complete breakdown. Gross notes that

> (...) only a specific class of perturbations will be able to affect an organism in such a way as to bring about a persisting state of disease. They must overcome the organism's mechanisms of defense and homeostasis, but at the same time not kill it right away. Conversely, it is obvious that states of disease themselves must exhibit a certain degree of robustness since they resist the organism's attempts to restore a healthy equilibrium. (Gross 2011, 481–2)

An organism is a dynamic system with several sets of states, each called a 'basin of attraction', which tends toward one stable state, trajectory or cycle, called its 'point attractor'. Healthy states are such point attractors, and so are diseases. Perturbation factors tend to push the organism out of its healthy 'basin of attraction', and once its resistance is overcome, it falls into another basin of attraction, this one possibly being pathological (Gross 2011, 484–5). Two examples are analyzed from that perspective, metabolic syndrome and cancer: it is interesting to note that this kind of view also pervades some groups of researchers (see Gross for references), not only philosophers of science. Ultimately, it is also noteworthy that this perspective explicates that many perturbation factors may lead to the same basin of attraction, which defines the same disease despite the variety of possible causes for such a transition. The question here is whether this health-disease transition itself is a necessary feature of organisms. It is probably the case that there are different basins of attraction for one organism; but this does not explain why different basins of attraction must be called either disease or health.

A New Disease Template for Disease Sciences?

Boorse's classic view on disease assumes that physiology consists in a functional analysis of health resulting in a description of all the interlocking functions that typically secure survival and reproduction in a reference class of a species (Boorse 1977). In the same paper, Boorse suggested that this conception of health and disease provided an unchanged theoretical template for all new data so far:

> In general, there is clearly some plausibility in the claim that the history of medical theory is nothing but a record of progressive investigation of normal functioning on the organismic, organic, histologic, cellular, and biochemical levels of organization, and of the increasingly subtle kinds of pathology this investigation reveals. (Boorse 1977, 560)

---

[7] This seems to involve the Boorsean notion of a 'reference class'.

This suggests that the concept of disease has to date been unchanged, despite progress in medical knowledge. Wakefield would probably not agree, since he advocates the view that the notion of function has itself evolved, but nevertheless he does agree to a form of continuity of the notion as regards medicine, a complementary view he calls 'black box essentialism' (Wakefield 1999).

Yet contemporary medicine has undergone a revolution through the landslide of data that follows from going both molecular and epidemiological: '–omics' on the one hand, evidence-based medicine on the other, have provided so much more information on diseases and diseases mechanisms and processes that it is a rightful question whether *any* functional description is still able to enlighten what health and disease consist in.

Observations of the mechanisms of a function or a dysfunction are characterized by a level of 'resolution', defined not by the scale of the entities involved, but by how much detail or complexity is resorted to (Richardson and Stephan 2007). At some resolution, functions appear clearly: in diabetes, in metabolic syndrome, the function of insulin, the function of glucose, the function of specific receptors of muscle cells, and so on, are obvious. Boorse's contention seems to be that in principle, functional analysis can go deeper without changing its form, maybe until it "bottoms out" in physical laws (Machamer et al. 2000). In fact, biology seems to be shifting already from a kind of description where attributing functions is what clarifies phenomena the most, to a kind of description where describing interactions between entities in mechanisms without function attribution is our best approach. The reason is that it is not very enlightening when complex processes are described, appealing to dozens of agents achieving dozens of effects, none of which are either necessary or sufficient, as is the case for metabolic processes. Should we say that the functions of a protein x_1 are to y_1, y_2..., *and* y_n, and that a function of x_1, x_2, ..., and x_n is to y_1? It is still possible in principle, but not very useful in fact. As Hartwell et al. put it,

a discrete biological function can only rarely be attributed to an individual molecule, in the sense that the main purpose of haemoglobin is to transport gas molecules in the bloodstream. In contrast, most biological functions arise from interactions among many components. For example, in the signal transduction system in yeast that converts the detection of a pheromone into the act of mating, there is no single protein responsible for amplifying the input signal provided by the pheromone molecule. (Hartwell et al. 1999)

For that reason, this shift, if it is indeed happening, would not be a radical change: mechanistic models are still interpretable as functional models (Piccinini and Craver 2011). The kind of data that '–omics' provide even suggests that functional analysis might not "bottom out" after all, but rather dissolve out in biochemical networks.

This obviously might have important consequences for the notion of disease: if it is not relevant anymore to describe causal roles as functions, what about dysfunctions? Can dysfunctional systems be described without any reference to attributable functions? These are the stakes.

For instance, Gross' attractor perspective does not resort to the notion of function – but at the same time, it makes no distinction between healthy and pathological states of a system. In network biology, the cell's activity is considered to consist in a "network of networks" (Barabási and Oltvai 2004). A crucial notion is that of a 'functional module', that is,

> a discrete entity whose function is separable from those of other modules. This separation depends on chemical isolation, which can originate from spatial localization or from chemical specificity. (Hartwell et al. 1999)

It is debated though whether such modules are 'real', the alternative being modular organization or a large integrated intracellular network (Ravasz et al. 2002). Even in the modularist view, functions could be read off from subparts of organisms down to the level of modules only, but not further down. On the other hand, it is a fact that entities intervening in one functional module can, and often do, intervene in another one. There are already interesting interpretations of what diseases might consist of in the light of the hypothesis of 'functional modules' (Debret et al. 2011). Down to the level of functional modules, it is possible to talk about dysfunctions. In the case of genetic disorders,

> A disorder then represents the perturbation or breakdown of a specific functional module caused by variation in one or more of the components producing recognizable developmental and/or physiological abnormalities. (Goh et al. 2007, 8688)

The result is that either the results of different genetic abnormalities are given the same disease label because they involve the same dysfunction via the involvement of various mechanisms, or they consist in different diseases although the dysfunction is the same. Either way, the notion of a disease as a dysfunction might be considered an approximation, and it is indeed a question, not a principle, that it is and will always be the appropriate framework to define what disease consists in.

This should suffice as a sketch of what awaits philosophical elaboration of the concept of disease beyond conceptual analysis.

Conclusion

The distinction between a pre-naturalized and a naturalized meaning of 'disease' casts some light on the limits of what philosophers of medicine have been focused on, namely conceptual or meaning analysis. It also shows why this traditional and central area of philosophy of medicine seems so unconcerned by important and relevant developments of philosophy of biology about the concept of disease.

Naturalization of the concept of disease in the form of a general theory of disease is a project, not a result. It supposes that philosophers of medicine stop being directly concerned about the boundaries between health and disease. Although the concept of disease is fuzzy, it nevertheless clearly excludes conditions or processes such as pregnancy, menstruation, ageing or being dead. From the point of view of conceptual analysis, it is therefore a mistake to consider one of these to be a disease. It is not necessarily so for a naturalized definition of disease. This is probably why some philosophers have stuck with conceptual analysis as a means against over-medicalization, thereby defined as undue medical treatment of a problem relative to our traditional concept of disease (plus or minus a few special cases such as birth control). Yet considering that, say, hyperactivity or sadness is indeed a disease does not imply that it should be treated medically, for the fact that there is a treatment is far from involving that the treatment is desirable.

Naturalization also supposes that philosophers of medicine delve deeper into cutting-edge research in biology and biomedicine. Instead of refining the possible meaning of 'function' and 'dysfunction' by confronting them to elementary examples or even imaginary ones, they would examine the strengths and weaknesses of functional explanations in the latest hypotheses and results of science, and consider which notions of 'function' or 'dysfunction' are required, and whether they are.

The ultimate goal of such a method is not a social one – that of protecting people against over-medicalization, for instance – but a scientific one.

It has often been pointed out that medical science has made 'tremendous' progress without any general theory of disease (Hesslow 1993; Kincaid 2008), maybe not even a consistent theory of one type of disease such as cancer (Kincaid 2008). Yet it does not follow that producing one would be useless. As a matter of fact, much had been achieved in various areas such as astronomy, clock-making or ballistics before Galileo, Huygens and Newton. It did not follow that a general theory of movement was useless. Similarly, the hope is that a general theory of disease would provide the means to understand it even better than we currently do, and provide a much more relevant concept of what disease is.

Acknowledgments Thanks to Marie Darrason, Elodie Giroux, Maria Gonzales, Juan Carlos Hernandez Clemente, Cristian Saborido Alejandro, Kathryn Tabb, David Teira, Philippe Huneman and anonymous referees for their helpful comments.

References

Achinstein P (1968) Concepts of science. Johns Hopkins Press, Baltimore

Barabási A-L, Oltvai ZN (2004) Network biology: understanding the cell's functional organization. Nat Rev Genet 5:101–113. doi:10.1038/nrg1272

Barabási A-L, Gulbahce N, Loscalzo J (2011) Network medicine: a network-based approach to human disease. Nat Rev Genet 12:56–68. doi:10.1038/nrg2918

Boorse C (1977) Health as a theoretical concept. Philos Sci 44:542–573

Boorse C (1997) A rebuttal on health. In: Humber JM, Almeder RF (eds) What is disease? Humana Press, Totowa, pp 1–134

Boorse C (2011) Concepts of health and disease. In: Gifford F (ed) Philosophy of medicine. Elsevier, Amsterdam, pp 13–64

Canguilhem G (1991) The normal and the pathological, New edition. Zone Books, New York

Chavali S, Barrenas F, Kanduri K, Benson M (2010) Network properties of human disease genes with pleiotropic effects. BMC Syst Biol 4:78. doi:10.1186/1752-0509-4-78

Darrason M (2013) Unifying diseases from a genetic point of view: the example of the genetic theory of infectious diseases. Theor Med Bioeth 34(4):327–344

Debret G, Jung C, Hugot J-P et al (2011) Genetic susceptibility to a complex disease: the key role of functional redundancy. Hist Philos Life Sci 33:497–514

Dupré J (2011) Emerging sciences and new conceptions of disease; or, beyond the monogenomic differentiated cell lineage. Eur J Philos Sci 1:119–131

Enç B (1976) Reference of theoretical terms. Noûs 10:261–282

Engelhardt HT Jr (1975) The concepts of health and disease. In: Engelhardt HT Jr, Spicker SF (eds) Evaluation and explanation in the biomedical sciences. D. Reidel Publishing Company, Dordrecht, pp 125–141

Germain P-L (2012) Cancer cells and adaptive explanations. Biol Philos 27:785–810

Goh K-I, Cusick ME, Valle D et al (2007) The human disease network. Proc Natl Acad Sci USA 104:8685–8690. doi:10.1073/pnas.0701361104

Gross F (2011) What systems biology can tell us about disease. Hist Philos Life Sci 33:477–496

Hartwell LH, Hopfield JJ, Leibler S, Murray AW (1999) From molecular to modular cell biology. Nature 402:C47–C52. doi:10.1038/35011540

Hempel CG (1952) Fundamentals of concept formation in empirical science. University of Chicago Press, Chicago

Hempel CG (1977) Formulation and formalization of scientific theories. In: Suppe F (ed) The structure of scientific theories. University of Illinois Press, Chicago, pp 244–265

Hesslow G (1993) Do we need a concept of disease? Theor Med Bioeth 14:1–14

Horwitz AV, Wakefield JC (2007) The loss of sadness: how psychiatry transformed normal sorrow into depressive disorder, 1st edn. Oxford University Press, New York

Jeong H, Mason SP, Barabási A-L, Oltvai ZN (2001) Lethality and centrality in protein networks. Nature 411:41–42. doi:10.1038/35075138

Jimenez-Sanchez G, Childs B, Valle D (2001) Human disease genes. Nature 409:853–855. doi:10.1038/35057050

Kincaid H (2008) Do we need theory to study disease?: lessons from cancer research and their implications for mental illness. Perspect Biol Med 51:367–378

Lemoine M (2009) The meaning of the opposition between the healthy and the pathological and its consequences. Med Health Care Philos 12:355–362. doi:10.1007/s11019-008-9163-x

Lemoine M (2013) Defining disease beyond conceptual analysis: an analysis of conceptual analysis in philosophy of medicine. Theor Med Bioeth 34(4):309–325

Lemoine M (forthcoming) Is the evolutionary component of Wakefield's "Harmful dysfunction analysis" stipulative? In Faucher L, Forest D (eds) Discussing the harmful dysfunction analysis of mental disorders. MIT Press, Cambridge, MA

Loscalzo J, Barabasi A-L (2011) Systems biology and the future of medicine. Wiley Interdiscip Rev Syst Biol Med 3:619–627. doi:10.1002/wsbm.144

Loscalzo J, Kohane I, Barabasi A-L (2007) Human disease classification in the postgenomic era: a complex systems approach to human pathobiology. Mol Syst Biol 3:124. doi:10.1038/msb4100163

Machamer PK, Darden L, Craver CF (2000) Thinking about mechanisms. Philos Sci 67:1–25

Millikan RG (1989) In defense of proper functions. Philos Sci 56:288–302

Moghaddam-Taaheri S (2011) Understanding pathology in the context of physiological mechanisms: the practicality of a broken-normal view. Biol Philos 26:603–611

Mossio M, Moreno A (2010) Organisational closure in biological organisms. Hist Philos Life Sci 32:269–288

Mossio M, Saborido C, Moreno A (2009) An organizational account of biological functions. Br J Philos Sci 60:813–841

Murphy D (2006) Psychiatry in the scientific image. MIT Press, Cambridge, MA

Murphy D, Woolfolk RL (2000) Conceptual analysis versus scientific understanding: an assessment of Wakefield's folk psychiatry. Philos Psychiatry Psychol 7:271–293

Neander K (1991) Functions as selected effects: the conceptual analyst's defense. Philos Sci 58:168–184

Nervi M (2010) Mechanisms, malfunctions and explanation in medicine. Biol Philos 25:215–228

Nesse RM (2001) On the difficulty of defining disease: a Darwinian perspective. Med Health Care Philos 4:37–46

Nordenfelt L (1995) On the nature of health: an action-theoretic approach. Kluwer Academic Publishers, Dordrecht

Park J, Lee D-S, Christakis NA, Barabási A-L (2009) The impact of cellular networks on disease comorbidity. Mol Syst Biol 5:262. doi:10.1038/msb.2009.16

Piccinini G, Craver C (2011) Integrating psychology and neuroscience: functional analyses as mechanism sketches. Synthese 183:283–311

Putnam H (1975) Mind, language, and reality. Cambridge University Press, Cambridge/New York

Quine WVO (1981) Theories and things. Harvard University Press, Cambridge, MA

Ravasz E, Somera AL, Mongru DA et al (2002) Hierarchical organization of modularity in metabolic networks. Science 297:1551–1555. doi:10.1126/science.1073374

Richardson RC, Stephan A (2007) Mechanism and mechanical explanation in systems biology. In: Boogerd F, Bruggeman FJ, Hofmeyr J-HS, Westerhoff HV (eds) Systems biology. Philosophical Foundations, Elsevier, Amsterdam, pp 123–144

Sadegh-Zadeh K (2000) Fuzzy health, illness, and disease. J Med Philos 25:605–638. doi:10.1076/0360-5310(200010)25:5;1-W;FT605

Severinsen M (2001) Principles behind definitions of diseases – a criticism of the principle of disease mechanism and the development of a pragmatic alternative. Theor Med Bioeth 22:319–336

Suppe F (1977) The search for philosophic understanding of scientific theories. In: Suppe F (ed) The structure of scientific theories. University of Illinois Press, Chicago, pp 3–232

Thagard P (2006) What is a medical theory? In: Paton R, McNamara LA (eds) Multidisciplinary approaches to theory in medicine. Elsevier, Amsterdam, pp 47–62

Thompson RP (2011) Models and theories in medicine. In: Gifford F (ed) Philosophy of medicine. Elsevier, Amsterdam, pp 115–136

Wakefield J (1992) The concept of mental disorder: on the boundary between biological facts and social values. Am Psychol 47:373–388

Wakefield J (1999) Mental disorder as a black box essentialist concept. J Abnorm Psychol 108:465–472

What Will Psychiatry Become?

Dominic Murphy

Abstract Modern psychiatry aims at uncovering the causal structure of mental illness. I discuss two issues relating to this. First, the allure of reductionism, which goes along with a metaphysical commitment to levels of explanation that gets in the way of more promising approaches to psychiatric explanation. Second, I discuss the place of psychology within psychiatry, suggesting that we may need to develop new psychological concepts to do justice to neuroscientific developments, but that this might rob psychiatry of the ability to help patients understand themselves.

Introduction

This paper is about the kind of science that psychiatry needs, and a plea to shake off a way of thinking that suggests one popular answer to that question. Many scholars seem to think that what psychiatry needs is genetics, and/or some sort of reductive neuroscience. While those approaches are definitely powerful, there is a lot they simply do not capture, and other thinkers have emphasized the need for psychiatry to employ many different scientific approaches altogether. I agree with the latter perspective, but I think that it can be misleading to put the point in terms of levels of explanation, as is often done. Levels of explanation are often either different ways of describing the same system, or else another way of talking about levels of constitution, with smaller units making up bigger ones within the hierarchy of nature that runs from atoms to organisms. A natural corollary of this hierarchical picture of levels is a reductionist agenda that concentrates on lower levels in the hierarchy. I will call some-times call this a "levels-based" approach. For various reasons, I don't think this suits psychiatry. I will say why, and investigate some of the consequences.

D. Murphy (✉)
Department of Philosophy, University of Sydney, Sydney, Australia
e-mail: Dominic.murphy@sydney.edu.au

P. Huneman et al. (eds.), *Classification, Disease and Evidence*, History, Philosophy and Theory of the Life Sciences 7, DOI 10.1007/978-94-017-8887-8_3,
© Springer Science+Business Media Dordrecht 2015

Whatever particular sciences we expect to find contributing to psychiatry, there is a consensus, among biologically inclined thinkers, about its trajectory; it will go on to more fully "discover the facts about how things go wrong with the psychology and biology of human beings" (McNally 2011, p. 216). I read this is as a commitment to a strong version of the medical model in psychiatry. It says that humans are made up of biological systems with a natural function, and science can discern these functions and say when they are not being discharged as nature intends. I have distinguished (Murphy 2013) this strong interpretation of the medical model from a weaker version. The weak interpretation is committed to gathering information about signs, symptoms, risk factors, treatments etc., but lacks the ontological commitments of the strong version, which sees mental disorders as pathologies of mechanisms of the nervous system. The weak version sees mental disorders as syndromes and is agnostic about their biological basis.

This essay will be concerned with the strong version of the medical model. It assumes that we can talk about some brains as dysfunctional and others as falling within biologically healthy ranges of functioning, and that the explanation of the abnormalities can be carried out using the vocabulary of some favoured sciences of the brain. After setting this out in a little more detail, I'll ask what those sciences might be. Despite some recent trends in psychiatry, I shall argue that nothing in the medical model requires us to restrict ourselves to purely biochemical explanations. Rather, I will argue that the logic of causal explanations in psychiatry make many types of explanation possible. This position makes reductionism less attractive, but it also raises questions about the sorts of cognitive theories that we might employ.

The Medical Model

The strong version of the medical model (and from now on I'll just say "medical model") interprets mental illnesses the way biological disease has been seen since the nineteenth century; to wit, as departures from normal functioning in some biological system. So understanding disease means understanding the normal function of bodily systems, and in psychiatry that means the brain or the central nervous system. Mental disorders are realized in neurological systems that are not doing what they should. So we need to understand normal and pathological function in terms that make that failure perspicuous – we ask what the system has evolved to contribute to the overall system that it partly constitutes, and how it is failing to do that. As Thagard (2008, p. 340) puts it clearly:

> the circulatory system consists of a set of components—the heart, veins, arteries, and blood—that interact to provide nutrients to the rest of the body. This mechanism is susceptible to many kinds of breakdown, such as defects in the heart valves, blockage in the arteries due to plaque and blood clots, and abnormal growth of blood cells. These breakdowns can arise because of many kinds of interacting causal factors, from internal ones such as defective genes to external ones such as infectious agents.

Similarly, the explanation of mental diseases requires specification of the normal functioning of the brain and other relevant organs, along with precise description of the different kinds of breakdown that can impede mental functioning.

This sets out the agenda very clearly, but obviously raises numerous questions. The very idea that science can tell us what goes wrong with people in such a straightforward fashion is itself very controversial, since many theorists contend that specifying the correct functioning of a biological system is not a simply scientific question. Every species exhibits variation – this idea is the essence of Darwinism, so we should expect all biological systems to vary across members of a species. The question, then, is how to give sense to the idea that there is some correct state in which a natural system should remain, in the absence of final causes or some other way of saying what nature ought to be like.

Kincaid (2008, p. 375) has argued that it is unreasonable to see the understanding of the normal function of biological systems as part of the medical model. We can investigate depression (his example) based on "partial and unsystematic" understanding of its causes, as we do with organs in medicine more generally. But Kincaid identifies the possession of background theory with having a "complete wiring diagram of the organism from fertilization to maturity" (p. 377). Furthermore, he sees the search for such wiring diagrams as reflecting a view of science as a search for laws of nature and natural kinds. But these commitments do not have to hang together so tightly: a background theory of what a system does can be quite vague, but without some understanding of a system's typical function it is difficult to see how we could reach the conclusion that there was something wrong with it.

Graham (2010, pp. 53–58) offers a different criticism of the medical model which turns on the (plainly correct) observation that most mental disorders are the product of several different causes rather than one exclusive one, or in his terms, a set of propensities rather than a "single main cause" (p. 55) as in bodily diseases like malaria. However, this objection can be met if we distinguish between the realization of a disease and its more proximate causes. Many diseases have a number of different possible causes that interact with genetic propensities; lung cancer is not just caused by smoking, for example, but also by inhaling various pollutants such as asbestos or coal dust. The different causes share the power to exert a destructive effect on the respiratory system via the replication of abnormal cells of various sorts. We lack a comparably detailed story for mental disorders, but the logic would be the same. The proponent of the strong medical model bets that the different causes of a mental disorder will tend to render a set of neurological systems abnormal in the same way across the affected population, even if the details of the cases vary according to accidents of biography. It is not in dispute that the subjective intentional life of the patient makes a difference to how mental illness is experienced and manifested in different people. The medical model's fundamental contention is that all these people, despite the varieties in their presentations, have something in common at the neurological level: their neuropsychological systems are disrupted in ways that we make sense of using the explanatory resources of the neurosciences, including cognitive or intentional concepts. The proponent of a levels-based approach further insists that all the causes that push the system into dysfunction must be amenable to micro-reduction. That might seem plausible if we think only of the relation between brain systems and their components, but it will not work when we widen our focus to take in the other, more distal, causes.

A second complication that Thagard's passage raises is that of the relation between dysfunction and disease. It is generally agreed that even if science can tell us when a biological mechanism is not working properly, that alone does not justify calling someone diseased. The additional judgement requires a different sort of basis, and most scholars agree that it must come from the norms of the surrounding community. This judgement is easy to make and share in cases where extreme pain, suffering or risk of death are present. But in many other instances, including a lot of psychiatric cases, judgements are likely to be contested. For example, suppose you inform your doctor that you wish to have your right leg amputated below the knee. It's very common to assume that if you have that desire, you are basically crazy. But the desire shows up in people whose mental health is otherwise unimpeachable, and they sometimes suffer acutely from the presence of what they feel to be an extraneous and unsightly body part. Perhaps we should just treat them as we would treat someone who wants a face lift, breast alteration or substantial tattooing; as harbouring a desire for a dramatic bodily alteration that might be unusual, but is not evidence of mental disorder.

Judgements that somebody is sick, bodily or mentally, are a particular family of judgements of deviance. Since communal norms are so important to our existence humans are deviance-detecting animals, and we draw many distinctions among counternormative behaviour. Sometimes we call it criminal, sometimes it is seen as immoral or eccentric. It can even come to be admired, and if it is then it might shift norms in a new direction. Some norms are violated in a way that makes us see people as ill, and specifying exactly what those deviant phenomena are is a tricky business.

The interaction between these two problems – judgements of malfunction and judgements of deviance – lies at the conceptual core of philosophy of medicine. Many philosophers think that we can be objective about at least one of the two steps; science can tell us what has gone wrong with a person's biology, and then we can ask whether the effects of that dysfunction on a person's life are of the right sort, whether in nature or severity, to make a diagnosis of illness or disorder.

It may be that the right picture is the reverse of this. We make a judgement of illness and then look for a scientific legitimation of it by investigating the biology or psychology of the subject to find out what might be wrong. Thinking of the procedure in this second way suggests that it is our habit of policing deviance, not our attunement to dysfunction, that is driving the show. In either case, judgements of dysfunction are critical. The medical model, conjoined to the tradition that sees cognition as information-processing, places the cause of mental problems in the failure of neurological mechanisms to function as they should. I am going to set aside all the big questions about whether we can make judgements of natural function in the absence of final causes. We still face the question I want to explore in the rest of the essay. How can judgements of malfunction be made in a way that helps psychiatric explanation, classification and understanding?

A Tradition of Computational Neuroscience

The specification of the system of interest and the ways it can break down need to mention whatever concepts are necessary for understanding. The components of the brain are systems that govern the cognitive, sensory and motor capacities of the organism. It is normal in the neurosciences these days to view these systems as processing information. Cognitive scientists now employ computational models based on conceptions of information processing developed in the middle of the twentieth century. However, the basic idea of the central nervous system as a computational system dates back to the late nineteenth century and the idea that biological relations among parts of the nervous system can be modelled mathematically as dynamic transformations of the weights assigned to energy levels in and between cells, so that the output of a neuron is a function of the inputs to it. Nervous energy flowing through the system was modelled as sensory information ultimately derived from the environment, and specific states of at least sensory systems could be correlated with external states of affairs. Associations between ideas were modelled as changes in the connections between brain cells. William James (1890/1981, pp. 616–617) for example, offers a toy model of memory as a graph, with vertices interpreted as interconnected "nerve-centres" and retention depending on the strength of the connecting edges, which James calls 'paths' located "in the finest recesses of the brains tissues". This conception of the nervous system informs Freud's early thinking about the mind (in his posthumously published *Project for A Scientific Psychology* (Freud 1895)) and was common among his teachers (Glymour 1992; Wollheim 1990, ch. 3). The nineteenth century nervous system was a computational system in which mental activity is a process that adjusts connections between cells and the energy levels within them. We should distinguish this wider tradition from more recent claims which are characteristic of cognitive psychology, viz. that thought is manipulation of symbols: physical entities with semantic and syntactic properties.

Modern Computational Neuroscience and Psychiatry

The tradition I refer to predates modern theories of computation, but it can clearly be seen as a forerunner of connectionism. I have followed (Murphy 2006) a wing of the strongly medical psychiatric community in urging that psychiatry should adopt the methods of contemporary cognitive neuroscience, as they descend from the information processing tradition, in order to carry forward the research program contained in the ideas of the medical model, in which classification and causal explanation will be ultimately founded on the neurophysiological organization of the mind. This approach is only one way to apply the medical model, and makes a bet that cognitive neuroscience is able to account for psychological phenomena by treating them as computational processes (though not necessarily symbolic process, rather than connectionist ones). Skeptics about computational approaches to cognition can adopt other neuroscientific applications of the medical model. The worry is that those rival approaches lack the resources to deal with cognitive processes.

It is people, not parts of their brains, that are psychotic. But the explanation of why somebody is psychotic will cite problems with neurological mechanisms like the executive system in dorsolateral prefrontal cortex and its relations – perhaps, a failure of inhibition – with cognitive systems that have evolved to subserve thought less tethered to reality. An explanation in terms of the physiology of cognition does not rule out a broader range of upstream factors as sources of the functional disruption.

Following Kraepelin, we can distinguish etiology and pathology. An explanation of why Jane undergoes a psychotic episode could make reference to her recent trauma, or a failure to negotiate certain developmental challenges and a reliance on very destructive defence mechanisms (such as massive splitting and projection). To fit in with the logic of the medical model in its strong guise, however, such processes would need to have, among their effects, a realisation of a destructive or dysfunctional disease process in the brain. The ensuing neuropathology is just what the disease amounts to, on the strong interpretation of the medical model. That does not mean that the pathology must always arise in the same way, but if mental disorder is brain disease then there must be in every case a neuropathology – an abnormal state of a neurological mechanism – that realises the disease.

These mechanisms are cognitive systems involved in the regulation of social behaviour – I will say more later about what "cognitive" is likely to mean in psychiatric contexts. The systems are parts of larger biological systems – ultimately parts of organisms or even societies – and this leads naturally to a reductionist approach. Because biological systems can be described in many ways, they seem to cry out for a treatment in terms of levels of explanation; the same entity can be given a cognitive, computational or molecular interpretation, and since these are interpretations of the same thing the reductionist impulse has a clear opportunity. Fundamentally, it can seem, all the higher levels are just expressions of the lower. Let me now move to the contemporary scene, and try to weaken the grip of this connection between neuropsychology and the metaphysics of reduction.

Levels, Mechanisms and Reduction

An influential statement of this reductionist impulse comes in Oppenheim and Putnam's famous "Unity of Science as a Working Hypothesis" (1958). They argued that in principle, psychological laws could be reduced to statements about neurons, which could be reduced to claims about biochemistry which could be reduced to atomic physics, and thus we could have a successful "microreduction" of psychology to physics. A microreduction in their sense is the decomposition of the entities in the theory being reduced into the proper parts of the reducing theory. The hope, and bet, is that this reducing theory will be the theory of the very smallest bits of nature.

The Oppenheim-Putnam picture is a very powerful and natural portrayal of a vision of explanation tied to a vision of the world. There is of course a large and detailed philosophical literature on the ramifications of this pair of visions, but I will not go into all that here. I just want to draw attention to the grip that the overall picture, in whatever specific form, has exerted. The world is seen as a hierarchy of levels of entities, with small ones nested inside bigger ones, and ultimately explaining how things work involves showing how the higher levels emerge from the lower. As well as expressing the metaphysics that dominate modern science and philosophy, the picture also comports well with an idea of explanation as involving showing how things work – taking the bigger thing apart to reveal the workings within it- as well as suggesting how laws at one level can hold at other levels.

But figuring out what it takes to unify science is one thing, and explaining particular psychiatric phenomena is quite another. There is no reason to suppose that they must be explained reductively, if that means employing only the concepts of low level molecular neuroscience or genetics (Microreduction, which involves explaining in terms of the laws of basic physics, might be possible in principle but is still a fantasy). One ideal of explanation in science involves showing how things work in terms of their components. This ideal naturally fits psychiatry and other biomedical contexts, because there is a dearth of laws in those contexts. So the mechanistic picture can be seen as validating the Oppenheim-Putnam vision if we forego laws and think about levels of mechanisms. On this account, what a system does is explained by the operation of the smaller systems that it comprises. But the explanation can work without being embedded in the Oppenheim-Putnam picture of the world at all. We can just think of explanation as depending on showing the processes that cause a phenomenon of interest to happen. Those factors do not have to be "lower": they just have to show us what happened. The ideal of explanation is showing how some things make something else happen, but the things that do the explaining do not have to be parts of the explanandum, nor from lower levels of the natural hierarchy.

Psychiatric phenomena should be explained using whatever concepts are necessary to explain them, and nothing in the logic of the medical model rules that out. It is true that contemporary cognitive science assumes that the mind decomposes into components and shows how the components work in concert to produce behaviour. It is also undoubtedly true, though, that mental illnesses are complicated phenomena; they are mixtures of behavioural, psychological and physical signs and symptoms which appear to depend on many different causes. The same condition in different people also varies in length and severity. For example, your chance of suffering from major depression depends on many factors. Genes certainly make a difference, but so do factors like the extent of the child abuse you suffered, the state of your marriage and your history of substance abuse, as well as stressful environmental events like unemployment or bereavement (Kendler and Prescott 2006, p. 281). Reducing unemployment or bereavement, as Oppenheim and Putnam think of reduction, is a fantasy. But they can still play parts in a causal model.

The medical model need only talk about causes, without specifying what they must be in advance, let alone assuming that there will be a microreductive account of causes. Genetics and long-term unemployment may come together to explain why somebody is depressed without providing a reductive picture in Oppenheim and Putnam's sense. Depressive episodes are not made up of genes and units of unemployment, even if they depend upon them; the hierarchical picture of nature is not a good fit here, and the mechanistic picture works only in a very extended sense. But we can still explain something in terms of interacting natural phenomena. I will come back to this in a moment, but I should note the bigger problem for the Oppenheim-Putnam picture in psychiatry.

From Laws to Mechanisms

The real problem that psychiatry (and not it alone) poses for the hierarchical picture of levels is that we are not, in psychiatry, dealing with one phenomenon described in different ways. Underpinning Oppenheim and Putnam's account, and the conceptions of level-based explanations that descend from it, is the idea that the same natural process is in view at each level, so that we are always talking about the same process. The psychology is realised in the microphysics, so that really the psychology just is the microphysics, only at a very high level of abstraction. A similar assumption is built into Marr's (1982, pp. 24–5) distinction between three levels of explanation in cognitive science, which has had a huge impact in philosophy. Marr's highest level specifies the computational task accomplished by the system we are studying. This says what the system does, specified in terms of what it computes. The middle level describes the actual representations and algorithms that realize the computation. The lowest level tells us how brain tissue or other material substrate, such as the parts of a machine, can implement the algorithm. Building a machine to do what natural systems do is what Marr was really after.

Again, Marr's three levels are different representations of the same process, which in his research program was vision, understood as the construction of a 3D representation of the world from two-dimensional data. This picture tallies with the Oppenheim-Putnam one, and also Craver's (2007) picture of mechanistic explanation, which stresses causal relevance. Causal relevance in this sense tells you how it is that something happening at one level makes a difference at another level: it is because the lower-level system is a part of the higher-level system.

The worry is that this account will fail to do justice to the fact that in psychiatry distinct causal processes work on different levels but are also indicative of distinct phenomena that do not exhibit part-whole relations. Say you become depressed through the interaction of genetic load and sudden bereavement; the latter is not reducible to the former.

A mereological picture of levels does not have to be conjoined with a reductionist approach to explanation. At the same time as Oppenheim and Putnam, Kenneth Waltz had something very like a picture of levels of explanation in *Man, the State and War* (1959), only he called them "images". Waltz's problem was explaining why wars start, and he argued for three images each of which gave the machinery for a different explanation. One is human behaviour – wars start because people are aggressive. A second is the nature of polities – wars start because of the internal dynamics of states. A third is the nature of the state system – wars start because of the threats and incentives faced by nations in a system of actors with no overall control. Waltz did not have the metaphysical preoccupations of a philosopher, but there are clear hints of part-whole relations among his images: people constitute states, which in their turn make up the state system. However, there is much less of a reductionist tendency in Waltz – the explanations at the level of the state and the state-system are regarded as autonomous, and some room is given for each of them, even though the metaphysics of states is part-whole.

Waltz is an exception to the harmony I noted above between the picture of the world as a metaphysical hierarchy and that of explanation as reduction. He, like Oppenheim and Putnam, wrote at a time when all explanation was taken to depend on laws – you explained a phenomenon by showing how it is only to be expected, given the laws. You find this in the clinical literature of the time too: Cronbach and Meehl (1955) assumed in their account of validity that a theory in psychiatry needed to be a network of laws. But that picture has come under attack in recent years with the rise of mechanistic accounts of explanation (Bechtel and Richardson 1993; Craver 2007; Schaffner 1993; Tabery 2004). I want to turn now to the question of how the mechanistic account fits with the picture of levels of explanation.

For example, suppose we want to understand the mechanism by which neurotransmitters are released (Craver 2007, pp. 4–6). This involves finding answers to questions such as: why does depolarization of an axon terminal lead to neurotransmitter release, and why are neurotransmitters released in quanta? The answers involve pointing out various entities, including various intracellular molecules, and showing how their properties allow them to act. The entities interact with each other to give rise to the phenomena that we want to explain. An explanation with these features is mechanistic. In recent years philosophers have stressed the way in which explanation in many sciences, above all the biological and cognitive, depends on finding mechanisms (Bechtel and Richardson 1993; Craver 2007; Schaffner 1993; Tabery 2009). Rather than seeing explanation as a search for laws, we seek the parts within a system of which the structure and activities explain the phenomena produced by the system. Philosophers disagree over exactly how to characterize mechanisms, but it is agreed that mechanisms comprise (i) component parts that (ii) interact to give rise to the phenomena of interest. It is generally agreed that a mechanistic explanation shows how the parts and their interactions give rise to the phenomenon we want to explain.

The mechanistic picture also fits with a hierarchical or mereological understanding of nature. Mechanisms come in levels too, on the face of it: there are mechanisms in the cell that contribute to the larger systems that the cell is part of. Humans decompose into subsystems – reproductive, respiratory, cognitive – that decompose into organs, and it is easy to see these as levels of mechanism.

Central to Craver's account of mechanistic explanation, for instance, is causal relevance between phenomena at different levels of explanation (Craver 2002). Causal relevance is defined in terms of manipulability and intervention. Events at one level are causally relevant in so far as they make a difference at another level. Causal relevance depends on realization. Levels of explanation, on this account, are, as we have seen before, actually descriptions of the same processes at different levels of resolution. A delusion can be understood in personal terms as a psychotic episode in the life of an individual that depends on relations between different psychological processes in different brain systems. These in their turn are involve cells whose operations can be studied in terms of the systems that constitute them, and on down to the molecular level. On this account, explanation in neuroscience, as in biology more generally, involves describing mechanism(s) at each level in ways that make apparent the relationships between causally relevant variables at different levels (Woodward 2010). Showing the causal relations between levels lets us integrate models of phenomena drawn from different areas of neuroscience. Clinical data, imaging studies and other high-level psychological information ultimately need to be systematically related to models of low-level phenomena such as the effects of neurotransmitter activity.

But again, we need to be careful in thinking of the biological hierarchy as licensing traditional levels-based thinking, because psychiatry does not deal with different interpretations of just one causal process, and its models look inherently multi-level in the sense that causal processes involve phenomena that span levels and are part of the same process, but are not just different ways of understanding one thing (Schaffner 1993, 2011 suggests that this is true throughout the life sciences). A complex causal structure at many levels is in a lot of ways a poor match for traditional level-based views, because the latter, I have suggested, has a natural tendency to reductionism.

Since the causes of many mental illnesses include a mix of genes and environmental factors we need to think about how environmental factors can be understood within the mechanistic program. These are different kinds of process, not different levels at which one process can be represented. If we were dealing with one process describable in different ways, then we could anticipate an integrative account in which higher-level variables get mapped on to lower-level ones. But even though it is hard enough to imagine a molecular or neurological reduction of a psychological construct it is even harder to imagine a reductive analysis of socio-cultural factors like unemployment or childhood sexual abuse. They have brain effects, but the brain effects vary across classes of individuals in ways that depend on other environmental and genetic contexts (see Kendler and Prescott 2006 for a comprehensive review.)

Appealing to levels of explanation is unobjectionable if it just involves a reminder that we need to relate variables of many sorts. But it is not clear that we have any principled grounds for sorting phenomena into levels, especially once we move beyond the organism: are unemployment and bereavement processes at different levels?

Marr did have a principled basis for distinguishing levels. He imagined them as descriptions of the same process (the construction of a 3D image from 2D retinal impacts) couched in the vocabularies of different sciences. But when we move outside the skull and begin introducing environmental factors and other kinds of cause, the Marrian picture looks less plausible. The topmost level in the

Putnam-Oppenheim picture, for example, is the social. But there is no one way to represent "the social" and even on their terms it is hard to see how it could be one level. In Waltz, for example, the state is at one level (or 'image') and the state system at another. This works because we have a straightforward part-whole relation between states and the international system, but consider religion. Is it a part of the social level, together with (say) family life and the economy, or are these all different? And if they are different, is the difference one of levels of explanation, or something else? Religious, economic and other social phenomena all have an effect on your health, so how do we represent them in a levels-type view, whether based on laws or on mechanisms?

I have tried to identify two problems for a view that allies level-based thinking with reductionistic thinking. One is the fact that the reductionistic approach is a poor fit with higher-level phenomena that are autonomous, rather than just an abstract description of the micro level. Another is that in psychiatry (though not just there) we want to deal with social phenomena, and lack any clear criteria for applying levels talk in that context.

Two Responses

One way to deal with these problems is to assume that we can ignore the outside world because information about it is represented in the brain. Then, if we can understand the process of information transmission in the brain, we can reduce that to the micro-level. Adolphs (2010) for example, assumes that social neuroscience begins with the transduction of social information, so that social factors are relevant only in so far as the system represents them. The mechanistic-reductionist program can go ahead. Methodological solipsism of this type will work as an explanatory strategy if we want to preserve the mechanistic understanding of the biological hierarchy in an enduring system. It will uncover proximate mechanisms and the causal relevance relations between them. It will not help us to isolate the relevant environmental factors and understand their effect on the organism (because we have to know what they are first to make the solipsistic strategy work).

A different option preserves much of the mechanistic approach but takes a different view of causal relevance and a more relaxed view of levels. Campbell (2008) argues for an interventionist approach to causation. This is the view that when we say X is a cause of Y we are saying that intervening on X is a way of intervening on Y (Woodward and Hitchcock 2003; Woodward 2003; Pearl 2000): manipulating one variable makes a difference to another. This is not a definition or analysis of causation in other terms, since it makes use of causal ideas – it just states that questions about whether X causes Y are questions about what would happen to Y if we did something to change X. Kendler and Campbell (2009) have argued that an interventionist model provides a rigorous way of articulating the idea that any combination of variables might characterize the causes of a disorder, whilst at the same time providing a clear test of what variables are actually involved, thus avoiding a simple-minded holism that just says that lots of things are relevant. Kendler and Campbell advance a picture of psychiatric explanation that looks for control variables that make a difference to behaviour, such as humiliation or genetic factors. But they do not expect to fit all the variables into a natural hierarchy in which events at one level are reductions of events at a higher level. Indeed, their picture, like that of the theory of causation it draws on, is silent about metaphysics. The point is that some set of variables can serve as what Campbell (2008, p. 209) calls the "control panel" for the system; there are some variables whose manipulation has a large effect on the outcomes. The moral of this tradition is that although correlation does not equal causation, patterns of correlations do. Or rather, by manipulating some variables we can have systematic effects on the phenomena, and this justifies using causal language (as when, based on correlations, we say that smoking causes cancer) and offers us opportunities to intervene in the system.

On this picture, environmental processes are part of the overall explanatory system in their own right, not just qua representations in the hierarchy of brain systems. We can continue to look for causal stories at many levels of explanation in the brain, but we do not have to face the worry of reducing environmental factors. On Kendler and Campbell's story, unemployment is a genuine cause of depression in so far as it makes a systematic difference to depressed patients, even if there is no explaining unemployment in terms of a mechanism. It is a genuine cause of depression in virtue of its difference-making properties.

Those difference-making properties cross levels. Depression counts as a cause of something neurological in its own right, and not just in so far as it is mediated by mechanisms that realise it. That is, cause and effect are related across levels. Kendler and Campbell contend (2009, p. 997) that interventionism "permits the clear separation of causal effects from the mechanistic instantiations of those effects", thus directly confuting the approach favoured by Craver and Bechtel (2007, p. 554) who argue that it stretches the concept of causation to breaking point to admit interlevel causes: they say that "to accept interlevel relationships as causal violates many of the central ideas associated with the concept of causation". Craver and Bechtel argue that we explain effects in terms of interlocking parts, and the relation across levels, they affirm, is one of constitution, not causation; causation can only be intra-level. Events at a level cause subsequent phenomena at that level. They in their turn realise higher-level phenomena. Interlevel causation, on this view, amounts to something causing itself, because different levels are different ways of talking about the same thing. Craver and Bechtel take causal relevance to be the relation borne by phenomena at one level to the lower-level phenomena they depend on, but causal relevance is not causation (Note that this notion of causal relevance is not the idea that something is a partial cause of a phenomenon (for which, see Northcott 2012)). The dispute here turns in part, then, on philosophical views about the nature of causation. Campbell (2008) argues on broadly Humean grounds that that we simply cannot tell in advance of inquiry what causal relations obtain in nature. We simply have to take our causal relations where we find them, including interlevel ones. I think Campbell (p. 214) is correct to see a commitment among reductionists to a view of causation that requires physical contact among cause and effect. This is far easier to imagine in biological processes than in the relations between psychology and unemployment.

Appealing to levels of explanation in psychiatry, then, can either be a reminder that we need to relate variables of many sorts in explaining the causes of disorder. Or it can represent a commitment to a seeing psychiatric explanation in terms of a biological hierarchy, with systems built up out of other systems. The debate over how to understand mechanisms and processes at different levels is partly an empirical one and partly bound up with philosophical views on reduction, explanation and causation.

The problem with Craver and Bechtel's view is that it is so natural to talk of cross-level causation in psychiatry. It really does seem to be the case that variables at different levels have an effect on each other. Their contention is that when this happens we are only entitled to talk about genuine causation when we have a mediation of the higher-level by a lower one. That is, it is not the state of being unemployed that interacts with your inherited depressive genes to make your mood pathological. Rather, what makes the difference is the physical basis of being unemployed.

There is obviously something correct about this metaphysically. On any naturalistic picture of the world, everything is ultimately physical, and so being unemployed is an ultimately physical phenomenon: being jobless must supervene on a complicated disjunction of states of the world that physics could describe. But it is just a fantasy to expect a molecular or microphysical theory of unemployment, and the restriction of proper causal language to intralevel relations looks unmotivated. To say that unemployment causes depression does not seem to violate any basic ideas about causation as far as I can see, even if it relates a social cause to a psychological effect. And as I noted above, in some cases we are simply not sure about how to identify levels: if I say that slavery caused the US Civil War, am I relating causes at one level or across two?

Let me say where I think the discussion has reached. I have argued that the medical model seeks causal explanations for mental illnesses, regarded as pathological states of neurobiological systems. This raises problems because of the diversity of the symptoms and causes that psychiatry seems to acknowledge. The question is how to explain and represent these pathologies and a natural way is to acknowledge that psychiatry involves multiple levels of explanation. However, the commitment to levels has emerged from a tradition that sees scientific explanation as a part of a bigger, reductive explanatory project. This picture may be metaphysically appealing but it does not fit psychiatry. It relies on the idea that we are dealing with a unique processes unfolding in a natural system which admits of description in several ways. But in psychiatry we are typically dealing with several processes that are describable in only one way. A part-whole reductionist materialism is fine as a philosophy of nature, but for the explanatory, epistemic projects of psychiatry it is of very little use. Instead, we should look for control variables, the manipulation of which have a robust effect on the system we are interested in.

So, my original question, what sort of science does psychiatry need, can be reframed as what family of control variables offer the most promise for intervention and manipulation in psychiatric contexts. The reductionist answer is – molecular ones. But that answer now appears as an empirical one. It can't be defended as a general metaphysical commitment, because that is beside the point, and on the face of it, it is not correct. Many non-molecular variables look to be just as good candidates for sites of manipulation and intervention. But as I said, this is now an empirical question, to be decided by measuring intellectual progress rather than fidelity to a picture of the world.

My bet, for what it is worth, is that all sorts of variables, from all sorts of sciences, will turn out to be relevant. But this answer raises another issue which I want to explore in the rest of the essay. The search for control variables is a search for scientific concepts – ways of representing natural phenomena that fit into our scientific apparatus for controlling the world. In psychiatry, part of the final family of concepts is very likely to be cognitive, or more broadly psychological. This can seem humane- a reaffirmation of human experience, of the mind, in the face of a reductionist agenda that is often charged with alienating the mentally ill by treating them as mere machines. However, this optimism is only feasible if the psychology we develop remains tethered to ordinary categories of thought. There is good reason to think this may not happen, and that the psychology we end up with will be heavily revisionist. It may therefore be just as remote and alienating as a purely biological psychiatry. I will end up taking up this issue, because answering the question, what sort of science does psychiatry need, involves answering the question of what psychology will become.

Psychology, Humanity and Science

The medical model tells us what our explanatory task is. We need to explain the observed causal-statistical network of signs and symptoms in a class of patients by identifying the mechanisms inside the organism and the external factors that affect them, either by developing a model of the natural hierarchy or (I have suggested) by developing a control panel of important variables. The reductionist impulse tells us that the explanatory theory we develop should draw on the resources of the very small – if not by employing microreductive concepts, then at least by employing molecular ones. But there is nothing in the logic of the medical model that requires explanations of any particular sort. The medical model enjoins us to search for the causes of mental illness, and those causes could be biological, cognitive, social, or anything else that enables us to explain and predict. The medical model is about establishing the right causal pathways, and this task can be done independently of any metaphysical commitment. An explanation does not need to be biological to be useful.

We can regard psychiatry, then, as a form of cognitive neuroscience. This formulation makes room for the existence of cognitive explanations; indeed, Bentall (2003) has argued that cognitive psychology will typically do more to explain psychotic symptoms than alternative approaches. However there are two complications that we need to address before we can resolve that cognitive psychology plays a role in psychiatry. The first is whether the nature of cognitive neuroscience leaves room for psychology at all as an explanatory programme. The second is whether, if psychology does exist within cognitive neuroscience, it will look anything like the psychology we use in everyday life, and whether, if it does not, the result will be alienating rather than liberating.

The research program of cognitive neuroscience assumes that the privileged decomposition of the human mind will be physiological. It looks for brain systems that have cognitive jobs to do, rather than for abstract computational systems. The old cognitive science (exemplified by Marr's approach) assumed that one could disentangle human psychology into abstractly described computational processes whose comprehension required no knowledge at all of the underlying biology. This picture has changed as we have learned, at least in the human case, that the decomposition of psychological capacities marches in step with the identification of physical structures within the brain that realize the component capacities. Typically, crucial evidence is provided by the absence of cognitive abilities in a subject who has an anatomical or physiological deficit in some brain area. The process is one that, in Glymour's words (1992) discovers "cognitive parts"; the ensuing decomposition is a physiology that is intentionally described. We use psychological language to characterise what it is that connected regions of the brain do – what makes them a system with a function.

An influential recent treatment of addiction by Ross et al. (2008), for example, starts from a set of behaviours that trouble any view of humans as fundamentally rational agents. For one thing, addicts "reverse preferences". That is, they expend resources trying to stay clean, and they also expend resources on their addiction. Rational choice theories suggest two reasons why you might behave like this. One is an increase in the relative value of short-term rewards as you approach them, so that they get more attractive the closer they are in time. Second, you might simply underestimate the costs of withdrawal. Both of these seem to be true of humans, which raises some puzzles. First; if these properties are shared by humans, why aren't we all addicted to all our preferred activities? A second puzzle is more specific; if anyone should be able to reliably estimate the costs of withdrawal it's former addicts, because they have been through it already. Yet addicts are more likely to get addicted than the general population. Ross et al. think that a cognitive psychological model is just the wrong sort of thing to do the explanatory job here, because physiology solves the puzzles. Specifically, they appeal to the operation of the dopaminergic system and its interaction with other systems.

The dopaminergic system is of interest to Ross et al. because it: learns environmental cues that predict reward; estimates comparative values of rewards; directs attention to cues that predict reward; prepares the system to act on those cues. The system, as they present it, has a set of functions that are described in intentional terms. My point is that this is a description, not an explanation, of how the system operates. The explanation mixes chemical and environmental concepts. The ventral tegmental area and Pars compacta of the substantia nigra release dopamine in response to surprising magnitudes or learned contingencies. This implements learning: a flood of dopamine (in nucleus accumbens) tells your reward system that whatever it was attending to was better than expected. This sets up a feedback loop to direct further attention and cue the motor cortex to take action. Ross et al. argue that these properties jointly predict a system that will be captured by unpredictable shifts in small magnitudes. What prevents widespread addiction even though we are all built like this is the existence of frontal and prefrontal circuits that inhibit impulsivity through the integration of cognition – which regulates input to the reward system and emotion – especially risk aversion.

The psychology in this theory describes the function of the systems. The real explanatory power comes from the nature of the dopamine system and its relationship to other systems. We can describe these systems in rough intentional terms- the (midbrain) learning system, the (frontal) executive system, but it is unclear whether we should regard these as explanatory at all, or just heuristics designed to convey the function of the physiological systems. The explanation in terms of physiology would seem to lose none of its force if the intentional language were removed, but would cease to exist if we stripped away the vocabulary of systems neuroscience.

So, although the logic of the medical model permits explanations using any concepts you like as long as they are useful, we have a case here in which the psychology seems to be playing second fiddle to the neuroscience. It may even be that the neuroscience will crowd out the psychology altogether, as some eliminitavists have suggested. Feyerabend (1963) raised questions about our capacity to reduce folk psychology to physiology, and argued that any successful materialist theory would undermine folk psychology, by showing that there was really nothing mental at all. On this picture, physiology will not reduce psychology so much as replace it, and we might wonder whether the sort of explanations anticipated by Ross et al. do the same, by showing us how an account that might have been put in psychological terms, to do with learning, habituation and impulse, can be reframed in purely neurological terms. Perhaps the sort of science that psychiatry needs will not include psychology at all, but not because psychology can be reduced to something else. Rather, psychology will just be shown to be explanatorily vacuous compared to the emerging neurosciences. I think this conclusion is premature, but I do think that the neurosciences will change psychology. They may do so in a way that has potentially serious implications for ordinary experience.

Some aspects of our psychology don't matter to us. If experts come and tell you that you don't know how your brain parses sentences or responds to pheromones, you might not be bothered. But other aspects matter a great deal – we all care about our memories, our emotional life or the sources of our behaviour, and we do not want to be told that we are systematically wrong about them, especially not if the truth is expressed in scientific language that is incomprehensible to us. The truth about fermentation might be hard to grasp, but it does not interfere with your drinking. The truth about love or belief might be more disquieting.

The more prominent eliminativists have questioned the scientific credentials of folk psychology on these grounds. Churchland (1981) did fret about the reducibility of folk psychology to neuroscience, but his scepticism centred on other criticisms. He argued that folk psychology could do nothing to explain many psychological matters, including mental illness. He also charged folk psychology with stagnation, since it had not changed since classical antiquity, and that it was poorly integrated with other sciences. These properties – explanatory poverty, stagnation and parochialism – are grave defects in a scientific theory. Churchland took them to be evidence that folk psychology was overdue for replacement by a successor theory, which he assumed would come from neuroscience. Like ether or phlogiston, beliefs and desires should be discarded as the non-existent posits of an obsolete theory.

Stich (1983) also argued that cognitive science would do without folk psychological concepts, assuming it would be largely computational and employ syntactically individuated states rather than intentionally characterised representations, and so not concerned with intentional states at all. Our folk concepts, he ventured, are too vague to be scientifically useful. Stich's point here is that there are many cases in which it is unclear whether the concept of belief really applies at all. Tamar Gendler has recently (2008) advocated for supplementing our notion of belief with one of *alief*. An alief is an automatic state that has some belieflike features, exerting some control over behaviour and cognition, and typically in tension with belief.

Hume considers the case of a man hung from a high tower in an iron cage. He 'cannot forbear trembling', despite being 'perfectly secure from falling, by the solidity of the iron which supports him' (Hume 1978; 1.3.13, p. 150). Hume puts this in terms of general rules (see Serjeantson 2005) learned from experience, with one rule supplied by imagination – that great height is dangerous – set against another, drawn from judgement – that iron supports are secure. Rather than judgement and experience, Gendler puts things in terms of alief and belief, which highlights the tension –does the man in the cage really expect to fall? No. But he can't help thinking it, or at least imagining it.

I suspect that the real terrain is more complicated than the simple tensions in Hume's or Gendler's accounts; there are probably lots of distinct information processing types in the brain that have some of the stereotypical aspects of belief. But we may very well need to draw a distinction between "bottom-up" processes that exert unreflective control, and "top-down" processes that are more deliberative and effortful. The eliminativist tradition may have been right to put to the potential revolution that science might work on our culturally bequeathed psychology, but wrong to think of abolition rather than reform.

Let's think again about the addiction example. I said that the story that Ross et al. tell is one in which there is very little psychology, but that might be because the psychology we currently have lacks the concepts to fit what the science has discovered. Gideon Yaffe (forthcoming) tackles this issue, asking what the dopamine signal actually represents. What within our commonsense repertoire of folk psychological concepts fits the activity of the dopamine system? All the science tells us is that dopamine represents some X such that we want more rather than less of X. But what is X? Should we think of the phenomenon in question as one of liking, wanting or valuing, for example? Addicts, past research suggests, can want something without much pleasure out of it (i.e. without liking it). Yaffe's suggestion is that the correct interpretation of dopamine signals is that they represent value. Addicts want drugs and this gives them a reason to value drug-getting at the time of consumption. But, as with the neuroscience of belief or the distinction between wanting and liking, we may find ourselves groping for new concepts as the neuroscience throws our traditional concepts into confusion. Scientific advances have often caused large-scale reforms of our culture's view of nature. The hard thing to accept is that it might have a similar effect on our view of ourselves. That would be an epistemic advance, but the worry, as I have said, is that these new vocabularies will deprive people of their ability to understand themselves by replacing a familiar vocabulary with a remote, scientific one. Psychiatrists will always need to be able to help people understand what they have become. The worry is that greater understanding of the mind will make it harder for us to explain people to themselves.

References

Adolphs R (2010) Conceptual challenges and directions for social neuroscience. Neuron 65:752–67

Bechtel W, Richardson R (1993) Discovering complexity. Princeton University Press, Princeton

Bentall R (2003) Madness explained. Penguin, London

Campbell J (2008) Causation in psychiatry. In: Kendler K, Parnas J (eds) Philosophical issues in psychiatry. Johns Hopkins University Press, Baltimore, pp 196–216

Churchland PM (1981) Eliminative materialism and the propositional attitudes. J Philos 78:67–90

Craver CF (2002) Interlevel experiments and multilevel mechanisms in the neuroscience of memory. Philos Sci Suppl 69:S83–S97

Craver CF (2007) Explaining the brain. Oxford University Press, New York

Craver CF, Bechtel W (2007) Top-down causation without top-down causes. Biol Philos 22:547–563

Cronbach LJ, Meehl PE (1955) Construct validity in psychological tests. Psychol Bull 52:281–302

Feyerabend P (1963) Mental events and the brain. J Philos 60:295–6

Freud S (1895) Project for a scientific psychology. In: Standard edition of the complete psychological works, vol I. Hogarth Press, 1953–74, London, pp 283–398

Gendler TS (2008) Alief and belief. J Philos 105:634–663

Glymour C (1992) Freud's androids. In: Neu J (ed) The Cambridge companion to Freud. Cambridge University Press, Cambridge, pp 44–85

Graham G (2010) The disordered mind. Routledge, New York

Hume D (1978) In: Nidditch PH (ed) A treatise of human nature. Oxford University Press, Oxford

James W (1890/1981) The principles of psychology. Harvard University Press, Cambridge, MA

Kendler K, Campbell J (2009) Interventionist causal models in psychiatry. Psychol Med 39:881–887

Kendler KS, Prescott CA (2006) Genes, environment, and psychopathology: understanding the causes of psychiatric and substance use disorders. The Guilford Press, New York

Kincaid H (2008) Do we need theory to study disease? Lessons from cancer research and their implications for mental illness. Perspect Biol Med 51:367–378

Marr D (1982) Vision. W.H. Freeman, San Francisco

McNally RJ (2011) What is mental illness? The Belknap Press, Cambridge, MA

Murphy D (2006) Psychiatry in the scientific image. MIT Press, Cambridge, MA

Murphy D (2013) The medical model and the philosophy of science. In: Gipps R, Fulford W (eds) Handbook of the philosophy of psychiatry. Oxford University Press, Oxford, pp 966–986

Northcott R (2012) Partial explanations in social science. In: Kincaid H (ed) Oxford handbook of philosophy of social science. Oxford University Press, Oxford, pp 130–153

Oppenheim P, Putnam H (1958) The unity of science as a working hypothesis. In: Feigl H et al (eds) Minnesota studies in the philosophy of science, vol 2. Minnesota University Press, Minneapolis

Pearl J (2000) Causality. Cambridge University Press, Cambridge, UK

Ross D, Sharp C, Vuchinich RE, Spurrett D (2008) Midbrain mutiny: the picoeconomics and neuroeconomics of disordered gambling. MIT Press, Cambridge, MA

Schaffner K (1993) Discovery and explanation in biology and medicine. University of Chicago Press, Chicago

Schaffner K (2011) A philosophical overview of validity. In: Kendler K, Parnas J (eds) Philosophical issues in psychiatry II: nosology. OUP, Oxford

Serjeantson R (2005) Hume's general rules and "The chief business of philosophers". In: Frasca-Spada M, Kail PJE (eds) Impressions of Hume. Oxford University Press, Oxford, pp 187–212

Stich S (1983) From folk psychology to cognitive science. MIT Press, Cambridge, MA

Tabery J (2004) Synthesizing activities and interactions in the concept of a mechanism. Philos Sci 71:1–15

Tabery J (2009) Difference mechanisms: explaining variation with mechanisms. Biol Philos 24:645–64

Thagard P (2008) Mental illness from the perspective of theoretical neuroscience. Perspect Biol Med 51:335–352

Waltz K (1959) Man, the state and war. Columbia University Press, New York

Wollheim R (1990) Freud. Cambridge University Press, Cambridge

Woodward J (2003) Making things happen. Oxford University Press, New York

Woodward J (2010) Causation in biology: stability, specificity, and the choice of levels of explanation. Biol Philos 25:287–318

Woodward J, Hitchcock C (2003) Explanatory generalizations, Part I: A counterfactual account. Nôus 37:1–24

Yaffe G (forthcoming) Are addicts akratic?: interpreting the neuroscience of reward. In: Levy N (ed) Addiction and self-control. Oxford University Press, Oxford

... ...

The Function Debate and the Concept of Mental Disorder

Steeves Demazeux

Abstract In this paper I compare the functional approaches of two authors, Christopher Boorse and Jerome Wakefield, and I focus specifically on the solutions that they offer to resolve conceptual difficulties in psychiatry. I demonstrate that their respective positions are ambiguous: the solutions they propose waver between two opposite points of view. The one is a denunciation of the psychiatric discourse from the perspective of what it *should* be; the other is a legitimization of what this discourse *is* in respect to the limited state of psychiatric knowledge. I argue that this vacillation stems from the fact that both authors, each in their own way, remain too indeterminate about the role of the concept of "biological function" in their definition of mental disorder. I seek to show that this frailty undermines the practical value of both Wakefield's and Boorse's analyses.

Introduction

In the mid-1970s, a renewed interest in the concept of function surfaced in two emerging disciplinary fields, the philosophy of biology and the philosophy of psychiatry. On the one hand, in philosophy of biology, Larry Wright proposed an etiological analysis of function. His account, first propounded in a 1973 paper that was to fuel the debate for decades to come, aimed at reducing the apparent problematic teleological characteristics of our functional attributions to an ordinary causal form of explanation (Wright 1973). In 1975, Robert Cummins proposed a "systemic" conception of function, which sought to be more general but also more faithful to the traditional usage of the notion by biologists and physicians. According to him,

S. Demazeux, Ph.D. (✉)
Laboratoire SPH, Université Bordeaux Montaigne, Pessac, France
e-mail: sdemazeux@gmail.com

P. Huneman et al. (eds.), *Classification, Disease and Evidence*, History,
Philosophy and Theory of the Life Sciences 7, DOI 10.1007/978-94-017-8887-8_4,
© Springer Science+Business Media Dordrecht 2015

a function refers to the causal contribution of an element within a system (Cummins 1975). On the other hand, in psychiatry, the notion of "mental illness" constitutes an old problematic notion. Since the 1960s, many philosophers and sociologists have denounced the excessive use of psychiatric labels to categorize all sorts of deviant behaviours. At this time, psychiatry as a medical discipline was in crisis, and one of the most burning questions was whether its conceptual foundations were legitimate, starting with the general concept of "mental illness" or "mental disorder". Some authors began to defend the idea that the core of the problem lies with the concept of dysfunction – and hence with that of function. In 1976, Christopher Boorse offered a detailed analysis of the concept of mental disease which he based on the concept of mental (dys)function. Later, in 1992, Wakefield followed with an analysis more explicitly based on an evolutionist conception of mental functions.

In this chapter, I provide an analysis of the respective functional approaches of these two authors in order to formulate a critical synthetic view. But contrary to most commentaries that have approached them from the perspective of biology or from a broad medical point of view, I focus on the *specific solutions* that they offer to psychiatry. Indeed, both Boorse and Wakefield clearly claimed to provide an analysis that could help psychiatry resolve its conceptual difficulties. But the solutions they respectively offer waver between two contradictory points of view. The one is a denunciation of the dominant conceptions in the psychiatric discourse from the perspective of what this discourse *should be*; the other is a legitimization of what this discourse *is* in respect to the limited state of psychiatric knowledge. I argue that their common vacillation stems from the fact that both authors, each in their own way, remain too indeterminate about the role of the concept of "biological function" in their definition of mental disorder. I demonstrate that this limitation undermines the practical value of both Wakefield's and Boorse's analyses.

The Project for a Conceptual Analysis of the Concept of Mental Illness

The Origins of a Controversy

Christopher Boorse's and Jerome Wakefield's contributions to the philosophy of medicine aim to be very general in scope. Their respective goal is ultimately to resolve the old problem of the distinction between the normal and the pathological *for all medical phenomena*. Yet the initial problem they were confronted with was precise and specific: addressing anti-psychiatric critics who, in a nutshell, see mental medicine as a usurped medicine. Boorse and Wakefield, in their respective work, first sought to show that psychiatry was not a usurped medicine, or at least that this need not necessarily be the case. Their main aim was to provide an objective definition (partially objective in the case of Wakefield) of mental disorder. The general scope of their work primarily stems from the conviction that the validity of the definition they offered of the pathological phenomenon in general should be considered as the best guarantee for its validity at the local psychiatric level. In other words, the two authors agreed from the start that "mental disease" should be defined as a particular type within the general "disease" category.[1]

In what follows I pay a close attention to these two authors' primary aim of addressing anti-psychiatry critics. While Wakefield still explicitly roots his work in a reflection on psychiatry, Boorse has been reproached for ultimately neglecting the specific problem he started from.[2] Yet in a rebuttal written in 1997, the author himself regretted how little attention critics of his theory had paid to its "major goal", which was indeed to propose a new analysis of mental health (Boorse 1997: 60). It is therefore important to remember the context of controversy that fuelled the works of the two authors. Boorse's first article on the question (Boorse 1975), the most cited one, explicitly addresses the anti-psychiatry debate at a time when Thomas Szasz (1920–2012), the famous and highly controversial American anti-psychiatrist, fired his first attacks on the psychiatric institution. His argument that mental illnesses are nothing but "myths" (Szasz 1960) just hit the nail on the head of the main conceptual issue. Notwithstanding the lampooning and often sweeping style of his polemical writing, Szasz did point out a genuine epistemological difficulty in the definition of the pathological phenomena in psychiatry. Boorse, in his article, cites him in the introduction, and agrees with him that the legitimacy of the extension of medical vocabulary to the sphere of mental health remains *an open question*, to say the least (Boorse 1975: 50). He also recognizes the "psychiatric turn" as a characteristic of the twentieth century, in the sense of a growing temptation to resolve social and moral problems by drawing on the medical paradigm. But the aim of Boorse's 1975 text, as well as those that followed in 1976 and 1977, is to provide a positive answer to this open question: the project of a truly "mental" medicine is sound as far as one is able to define a set of typical mental functions of the human species.

Wakefield's work comes later. His first articles dedicated to the definition of mental disorder were published in the early 1990s, in a less polemical – still far from being pacified – context for the psychiatric institution. Wakefield criticizes the definition underpinning the third edition of the *Diagnostic and Statistical Manual of Mental Disorders* (DSM-III) by the *American Psychiatric Association* for being ambiguous and inaccurate. He offers a "Harmful Dysfunction Analysis" (HDA) that seeks to identify the two fundamental poles of any pathological phenomenon. The one is factual, relying on the specific knowledge about a certain dysfunction, and

---

[1] For a discussion on this point, see what Neil Pickering calls the "likeness argument" (Pickering 2006: 14). Please note that while some authors draw distinctions between *disease*, *illness* and *disorder*, these distinctions will be of no use in my argumentation. In what follows, these three terms are considered as strictly synonymous.

[2] This reproach was made by William Fulford in an open letter to Christopher Boorse. Fulford regrets that Boorse has adopted the wrong perspective: it is not through a better understanding of *disease in general* that a better understanding of *psychiatric disease* may be achieved. On the contrary, it is through a better understanding of the complexity of what constitutes a *psychiatric disease* that the conception of *disease in general* should be reformed (Fulford 2001: 80–85).

the other evaluative, requiring that this dysfunction can be seen as harmful according to social norms. These two elements – the mental dysfunction on the one hand and the harmful element on the other – are thought to be the two necessary and sufficient conditions for legitimately identifying a "mental disorder". Like Boorse, Wakefield's primary aim is to oppose critics of psychiatrics who see "mental disorder" as nothing but a myth or, at best, as the expression of a *pure value judgement*. On the other hand, also like Boorse, he opposes the behaviourists, who readily embrace the normative dimension of the concept and do not hesitate to reduce "mental disorder" to a mere "behavioural disorder", even if that means reducing a pathological phenomenon to a problem of social adjustment. In other words, for both authors the concept of function plays a decisive role in distinguishing the normal from the pathological.

Two Modern Versions of the Broussais' Principle

Boorse and Wakefield are not the first to promote an objectivist general approach to the pathological phenomenon. The French philosopher Georges Canguilhem, in *The Normal and the Pathological*, coined "Broussais' principle" the idealistic project to define the pathological phenomenon in fully positive terms, especially by identifying it with the defect or excess of a given physiological phenomenon. Canguilhem borrowed the term from Auguste Comte who, in the 40th lesson of his *Course of Positive Philosophy*, had praised François Broussais (1772–1838) for being the first physician to clearly formulate this principle of an essential continuity between physiology and pathology: never, he wrote, has the fundamental relationship between pathology and physiology been conceived of so directly and satisfactorily (quoted by Canguilhem 1991: 47). Contrary to Comte, Canguilhem criticized "Broussais' principle" for missing what he thought was fundamental for medicine about the pathological phenomenon, i.e. the importance of the inherent "normativity" of the living being. Canguilhem raised two fundamental objections concerning Broussais' principle:

> The ambition to make pathology, and consequently therapeutics, completely scientific by simply making them derive from a previously established physiology would make sense only if, first, the normal could be defined in a purely objective way as a fact, and second, all the differences between the normal state and the pathological state could be expressed in quantitative terms, for only quantity can take into account both the homogeneity and variation. (Canguilhem 1991: 57)

What Canguilhem wanted to demonstrate was that neither Broussais nor Comte, nor Claude Bernard later on, managed to fulfil these two important requirements. Did Boorse and Wakefield, the two advocates of a modern version of Broussais' principle, do better? At the risk of taking some liberties with Canguilhem's text, let's try to reformulate and formalize these two requirements of the objectivist contract. These will serve as a backbone throughout my discussion.

We can achieve an objective characterization of the pathological phenomenon only if:

(i) one can formulate a purely factual definition of the normal;
(ii) the difference between the normal and the pathological can be translated in quantitative terms.

As is evident, condition (i) is a necessary condition to objectively define the pathological phenomenon. This means proposing a definition of the normal that does not confine it to a mere value judgement. The emphasis is put on the theoretical relevance of a *factual definition of the normal phenomenon*. Both Boorse and Wakefield sought to fulfil this requirement in the same manner. First, in method-ological terms: the two authors rely on *conceptual analysis* in order to clarify and highlight the factual dimension of the concept of disease, namely the reference to a certain dysfunction. Then, with regard to the results: both authors insist that it is the concept of *biological function* that will allow us to talk about the normal as a fact. The normality of a function can be described completely objectively, provided that we know what this function consists of. For example, to say of a heart that it functions *normally* can be understood in biological terms and, contrary to appearances, as a purely descriptive and factual judgement. This is what the first clause is about.

Condition (ii) is trickier to interpret. For Canguilhem, the difference between the normal and the pathological can be thought of either as qualitative (a difference of nature), or as quantitative (a difference of degree). While, according to him, the dynamic polarity of the living should lead us to think of this difference as qualitative,[3] the only way to think of it in objective terms would be by managing to express it "in quantitative terms, for only quantity can take into account both the homogeneity and variation" (Canguilhem 1991: 57). This is a theoretical argument that considers the conditions under which we could conceive of an identity in spite of difference, and a difference through the identity. Indeed, how can we think of pathology as following the general rules of physiology, while appealing to an objective difference between the two disciplines? At a more practical level, this condition (ii) can be seen as asking how one can *objectively identify* the pathological phenomenon among all the phenomena of the living. Whereas condition (i) addresses the issue of the theo-retical validity of the objectivist approach, condition (ii) is more focused on the conditions *for its practical application in the biological and medical sciences*.

Let us return to Boorse and Wakefield, and to psychiatry. The great value of their two approaches, I argue, lies in the interesting solution they provide to the problem of the conceptual legitimacy of the objectivist approach to psychiatry (i). In response to Szasz, who had built his attack on the important intuitive role of the concept of *lesion* in somatic medicine, the re-evaluation of the concept of *biological function* sheds light on the potentially structuring role it can have in medical discourse in general and in psychiatric discourse in particular.[4] This re-evaluation thus offers new grounds on which to defend the validity of the medical model inside psychiatry, and therefore to defend the legitimacy of psychiatry as a genuine branch of medical knowledge.

---

[3] Canguilhem supported the idea that "disease is still a norm of life but it is an inferior norm in the sense that it tolerates no deviation from the conditions in which it is valid, incapable as it is of changing itself into another norm" (Canguilhem 1991: 183).

[4] It is important to note in this respect that Boorse and Wakefield, by defending the autonomy of the psychological discourse in psychiatry through the concept of function, are in complete opposition to Broussais, who in *De l'irritation et de la folie* incessantly attacked "psychologists" – whom he saw as nothing but disguised spiritualists. But I am interested here in "Broussais' principle", as it has been characterized by Comte and investigated by Canguilhem.

It is however one thing to show that psychiatry *can* be a legitimate medical discipline taking care of medical conditions that truly deserve the term "mental disease". But it is quite another to claim from a mere theoretical point of view that the current psychiatric discourse – as it is used and practised – is well founded. To establish the *conceptual* legitimacy of a notion is not sufficient to demonstrate its current practical legitimacy. The history of science is full of these errors where apparent theoretical validity is used to justify distorted discourses and practices. In this respect, it is worth remembering that the anti-psychiatry critics, for the most part, have based their criticisms on the psychiatric discourse *as it presents itself in reality* rather than what it could or would like to be. Szasz, with his "transcendental" objection (in the Kantian sense that it challenges the very conditions of the possibility of a psychiatric discourse as a medical discourse), is the exception rather than the rule. Most of the time, critics of psychiatry not only denounce the insidious power exercised by the psychiatric institution but also attack specific diagnoses by calling into question their theoretical construction or the discriminating power they allegedly hold to distinguish between normal and pathological behaviours. In other words, what truly matters is not so much the problem of overall legitimacy (i) as that of the means of a specific distinction (ii): what is the evidence, what are the effective grounds for distinguishing a given normal behaviour from its pathological counterpart? What does psychiatry base itself on to carry out its nosological categorizations in practice?

The Philosophers' Attitude Towards the Psychiatry of Their Time

To address the demarcation problem between the normal and the pathological (ii), we need to distinguish between the way the issue is addressed through the psychiatric literature and the way, perhaps, that psychiatry should deal with this problem. Boorse and Wakefield clearly adopt the same general attitude towards this problem, even if, as we shall see, they each provide a significantly different answer. Their general attitude is marked by two recurrent features.

The first is cautiousness with respect to the institutionalized psychiatric discourse. On several occasions, Boorse carefully distances himself from the psychiatry of his time. He readily recognizes the excessive use of psychiatric labels, or at the very least a tendency in the profession to too easily qualify as pathological behaviours what society condemns (Boorse 1976: 75). The title of the 1976 article "What a theory of mental health should be" clearly shows that his aim is more to reform or clarify the psychiatric project than to defend an institutionalized discipline that, according to him, is not doing well. The conclusion of this article is eloquent, with the author sharing his impression that for psychiatry, "the time has now arrived when the clinical disciplines face a parting of the ways" (*Ibid.*: 81), and that it must take a step toward greater theoretical clarity and rigour, or else clearly renounce any reference to the normal and the pathological. However, Boorse immediately adds, this does not mean that one needs to radically challenge the current state of psychiatric knowledge: despite the controversies and confusion, the author can hardly imagine that clinicians as a whole are "too wide of the mark" (*Ibid.*). Wakefield, in his own work, shares the same attitude toward institutionalized psychiatry, which can be favourably qualified as *constructive criticism*. The author however shows resolutely more enthusiasm and engagement than Boorse with some specific debates in psychiatry. Wakefield is a prolific author, who believes in the validity of most diagnostic labels, but also intends to remain very sensitive to the normative powers that these labels both have and that they are subject to.[5] His work has looked at numerous contentious issues, ranging from personality disorders to the diagnosis of depression, and from anxiety disorders to sexual disorders. His position could be summed up as a mitigated defence of the DSM-III project (and its later versions), mixed with an often incisive critique of the different aspects of the classification's practical implementation.

The second common feature in both works is the emphasis on the fact that scientific psychiatry is still in its infancy. Like Claude Bernard who, during his time, regretted that somatic medicine was still "a science in its infancy" as opposed to the "constituted sciences" like physics or chemistry, but who saw no reason just to wait and see, as we often gain to experiment (Bernard 1984: 50), Boorse and Wakefield argue that the psychiatry of our time must offer solutions with the means at its disposal, until it is able to positively draw on the "psychology of mental functions" that it needs. Boorse highlights the "conceptual wealth" and the "practical flexibility" (Boorse 1997: 101) of his theory, arguing that "as soon as biological dysfunction is required for disorder, virtually all the BST's[6] benefits accrue in clarifying professional and social controversies and preventing political abuse of medical vocabulary" (*Ibid.*: 100). As for Wakefield, he constantly highlights "the considerable explanatory power of the HD analysis for understanding the distinction between mental disorder and other problematic mental conditions" (Wakefield 2007: 149), using the HDA several times to test the validity of the diagnostic criteria of certain mental disorders.

Oscillating between critical cautiousness and firm confidence in the psychiatric discourse of their time, Boorse and Wakefield make themselves advocates of a scientific medicine which is clearly at the service of patients rather than of the insidious and normalizing power of the psychiatric institution. Their common conviction is simply that science and conceptual clarity can resolve a number of contentious issues in psychiatry. This certainly sounds convincing. The only thing is: are they in position to do this?

---

[5] Note that Wakefield was a student of Michel Foucault at Berkeley. Although he rapidly distanced himself from the French philosopher's anti-naturalist positions, Foucault remains an important reference in his work.

[6] Bio-Statistical Theory: the label Boorse gave his theory. Cf. *infra*.

Boorse's Populationist Approach

Psychiatry, a Legitimate Theoretical Enterprise

Of the two philosophers, Boorse can be said to be the one who most conscientiously addresses the challenge set by Canguilhem: he seeks (i) to propose a purely factual definition of the normal, based on a value-free conception of function in biology; and he claims to be able (ii) to translate the difference between the normal and the pathological into quantitative terms, through the approach that he calls "Bio-Statistical Theory" (BST).

The first point (i) is the one to which philosophers have paid the most attention: can we *objectively* define the pathological phenomenon based on a factual account of a biological dysfunction? The solution offered by Boorse in 1975 is simple and elegant, but it raises many difficulties. In his 1997 "Rebuttal on Health", Boorse answers the many technical, medical and biological objections that have been addressed to him over a period of 20 years. It is however surprising to note that in this paper, which covers over a 100 pages, specific references to the psychiatric problem – if they were all grouped together – would not exceed ten pages.[7] Psychiatry as a genuine branch of medicine, which was one of the issues addressed in his work in the 1970s, has become completely secondary in his 1997 rebuttal.[8] Part of the reason for such a shift is that to prove the correctness of his analysis, Boorse has increasingly come to favour some striking and intuitive examples drawn from somatic medicine at the expense of the more ambiguous and polemical ones he could have drawn from psychiatry. Still, in light of the importance given to increasingly small, even niggling objections (Is the dysfunction of a single cell enough to make it a disease? What to think of dysfunctions that protect people against certain diseases like HbS haemoglobin for malaria? etc.), one can only regret that his priority was no longer to offer psychiatry the theoretical means of answering to its critics, but simply to defend his theory for its own sake.

Actually, psychiatry was one of the main issues that Boorse addressed in his three medical articles[9] published between 1975 and 1977. In this respect it is interesting to note that these three articles, published in different journals, all stemmed from a unique original long manuscript, which was divided up to fit the requirements of scientific publication (Boorse 1997: 101). In that context, the 1976 article "What a theory of mental health should be", which is often considered as a

---

[7] A substantial part of which is dedicated to a discussion with Engelhardt about masturbation in the nineteenth century. In 2012, Boorse gave some "Replies to Recent Critics", but the case of psychiatry is no longer discussed. The only remarks on psychiatry are dubitative concerning its current scientific status.

[8] Yet, most authors that Boorse debates with are directly involved in the field of psychiatry, e.g. Fulford, Reznek, Kendell, Klein, Scadding and Wakefield.

[9] "*On the distinction between disease and illness*", "*Health as a theoretical concept*" and "*What a theory of mental health should be*".

minor contribution, should have been the cornerstone of Boorse's philosophical reflection, insofar as it is in this paper that the author seeks to resolve the most important issue regarding the application of the BST theory to the mental sphere: how can we speak of "natural functions" in the field of mental health?

"There Can Only Be Mental Health if There Are Mental Functions"

The core of Boorse's argument in his 1976 paper "*What a theory of mental health should be*" consists in defending the idea that it is reasonable to conceive of a number of biological functions that the human mind fulfils. This does not require a formal definition of what constitutes a biological function. One just needs to assume that the psychological sciences have the capacity to define a set of psychological traits common to *Homo sapiens*, the *dys*-function of which produces characteristic clinical syndromes. This possibility, from a purely theoretical point of view, only presupposes two things: first, that some mental phenomena can be considered as genuine *causes* of action; second, that *typical* causal relations at the mental level, in the functioning of the human mind, can be identified. These are substantial philosophical assumptions, but they are enough to be accepted as a means of broadly characterizing "mental functions".

Concerning the question of causality between mental phenomena, Boorse supports the argument Davidson made in his famous 1963 article, "Actions, Reasons and Causes": there is no reason not to recognize the possibility of certain mental events acting as genuine causes of action. This point is important, as it implies that contrary to what some of its critics claim, Boorse's approach can in no way be reduced to a defence of what is commonly called "biological psychiatry". Boorse uses the concept of "biology" in the sense of a "science of life" that is sufficiently broad to encompass both physiology and psychology. He certainly does not seek to reduce biology to physical-chemical processes. In 1997, Boorse has firmly reasserted that he sees no reason to reject the possibility of certain mental diseases being *purely psychological*.

As for the second question, regarding *typical* causal links, Boorse refers to the works of Freud, Piaget and Chomsky: psychology does indeed seem to be able to define certain characteristic functions of the human mind. The only difficulty is not to neglect the plasticity of mental functions, which is probably greater than that of the body's physiological functioning. But Boorse's main idea revolves around his conviction that one can identify a biological regularity in particular mechanisms of the mind, whose dysfunction can be characteristic of various pathologies. He thus concludes:

> If certain types of mental processes perform standard functions in human behaviour, it is hard to see any obstacle to calling unnatural obstructions of these functions mental diseases, exactly as in the physiological case. So far the analogy between physical and mental health is unproblematic. (Boorse 1976: 64)

The author nevertheless remains very evasive concerning the reality and the number of these mental functions. What matters most here is to promote the theoretical relevance of his approach. In this respect, it is worth remembering that in his 1976 work Boorse sees Freud's psychoanalytic theory as offering "the best model [...] of what a theory of mental health should be" (*Ibid.*: 78). Boorse thus explains himself: "Formally speaking, psychoanalytic theory is the best account of mental health we have. It closely follows the physiological model by positing three mental substructures, the id, ego and superego, and assigning fixed functions to each" (*Ibid.*). This anecdote highlights at least one thing: the issue is epistemological and methodological before being a question of theoretical orientation for psychiatry.

Psychiatry, Between Body and Mind

Does Boorse's original approach, which distinguishes physical diseases from mental diseases, not amount to locking psychiatry into the old Cartesian dualism? The second point that Boorse seeks to demonstrate in his 1976 paper is that psychiatry as a medical discipline is not condemned to dualism, any more than it is condemned to being nothing but a failed neurology. Drawing on the works of the philosophy of mind of the 1970s, namely by Davidson, Putnam and Fodor, he tries to show that it is possible to allow a degree of autonomy to the mind without necessarily falling into Cartesian dualism. For example, the *Mind/Brain Identity Theory*, which Boorse sums up in broad terms, allows for all mental events to be thought of as physical events, without each particular mental event strictly correlating with a particular physical event. This form of refined materialism reconciles the fact that "to want to eat a lobster on the spot" is a *mental event* that necessarily has a particular cerebral trace, with the fact that it will not necessarily be identified with a particular type of *neural event*. In other words, it is possible to both adopt a materialist approach and defend a view of epistemological duality of the mental and neurological sciences. At the time of writing his article, Identity Theory (which was often identified to "Functionalism") had raised considerable interest amongst philosophers. Boorse wished to emphasize the importance of these new considerations stemming from the philosophy of mind for psychiatry: "My purpose in rehearsing this view is to show why it is in no way obvious that psychiatrists who reject Cartesian dualism thereby destroy the autonomy of their discipline" (*Ibid.*: 66).

In fact, Functionalism in philosophy of mind offers a reply to one of the strong objections raised by Szasz against psychiatry: it is possible to speak of "mental diseases" even when no neurochemical dysfunction or anatomical lesion of the brain can be observed. The existence of a characteristic lesion in the brain no longer needs to be demonstrated to *legitimately* be able to speak of a mental disease. Psychiatry can lay claim to its own area of investigation, independently of neurology, and therefore make full use of the medical vocabulary.

Psychiatry therefore constitutes a sound theoretical project, and neither Cartesian dualism nor materialism is a metaphysical option that could potentially undermine the autonomy of this project. The only decisive question is whether we can speak of *causality* for mental phenomena and of *mental* function. If we agree on these two points, then we can consider the possibility of a science of the normal and abnormal functioning of the human mind, even though it still needs to be developed. In these general terms, Boorse does live up to the challenge, though many theoretical issues remain.[10]

But what is the specific issue for psychiatry? Fundamentally, it is to defend the idea that only a science of the normal and abnormal functioning of the human mind can provide a criterion for the distinction between the normal and the pathological. According to Boorse, clinical medicine is not able to provide such a criterion. At best it provides only a clue, and Boorse firmly maintains that science and theoretical medicine must take precedence over clinical medicine in defining the normal and the pathological (Boorse 1997: 48). In this perspective, it seems that Boorse would not hesitate to paraphrase French physician René Leriche (1879–1955): "If one wants to define disease, it must be dehumanized" (cited by Canguilhem 1991: 92). The only question is: *how* can we accomplish that? What can be done in psychiatry if we cannot rely on a definite science of the normal functioning of the human mind? Though the general framework of such a science would be offered, how are the limits between normal function and dysfunction at the mental level to be identified?

The Delimitation Issue: The Bio-statistical Approach

In a certain way, Christopher Boorse has been able to restore the theoretical relevance for the old project of a scientific psychiatry in the light of contemporary debates in philosophy in mind. But one important issue remains, however: are we able to translate the difference between the normal and the pathological into quantitative terms (condition ii)? Boorse addresses this issue from a general point of view in his 1977 article, "Health as a Theoretical Concept", but his most important arguments are to be found in his 1997 synthesis. To pinpoint his theoretical position, he accepts the label of "bio-statistical theory" used by philosopher Lennart Nordenfelt to emphasize the two key elements of his conception: the idea of biological function and the idea of statistical normality. The significance of the role that these two ideas play in the BST is captured in Boorse's general definition of health and illness:

1. *The reference class* is a natural class of organisms of uniform functional design; specifically, an age group of a sex of a species.
2. A *normal function* of a part or process within members of the reference class is a statistically typical contribution by it to their individual survival and reproduction.
3. A *disease* is a type of internal state which is either an impairment of normal functional ability, i.e. a reduction of one or more functional abilities below typical efficiency, or a limitation on functional ability caused by environmental agents.
4. *Health* is the absence of disease. (Boorse 1997: 7–8)

---

[10] Fulford, for instance, highlighted that Boorse's conception of a biological function (as a subsystem's contribution) may not be relevant in psychiatry, where most disorders are related to the functioning of central mental features, like rationality.

The key point here is the mention of "statistically typical contribution": Boorse seeks to translate biological norms into the language of statistics. In this sense, we could say that the originality of his bio-statistical approach lies in a certain way of reconciling Claude Bernard's physiology with Adolphe Quételet's conception of the average human being (Giroux 2009). But the difficulty is that one cannot speak of the "normal function" of an organ or process in biological terms without presupposing a certain "species design", that is, the rather uniform characteristics shared by all members of the same species. Yet, from 1977 Boorse recognized the importance of not slipping into an essentialist conception of the functional properties shared by a single species, and consequently reifying the *design* of the species and neglecting the biological variability of phenomena. His definition of the normal function as a "statistically typical contribution" of a trait for survival or reproduction of the individual provides an answer to those who accuse him of essentialism. In this respect, Boorse draws on Ernst Mayr's classical opposition between typological and populationist approaches in biology and he claims that "BST is indisputably populationist" (Boorse 1997: 39). Far from giving into an essentialist (or typological) way of thinking, his theory highlights the fact that the *design* of a species is nothing but a precarious "statistical abstraction" (Boorse 1977: 558). And the same applies, albeit on another scale, to normal and pathological phenomena. In his 1997 paper, Boorse provides a bell-curve representation of his bio-statistical theory (Boorse 1997: 8).[11] For a given part-function, a pathological state can be statistically defined when the efficiency of the function is below a certain range.

But how should we interpret the condition (ii) of the objectivist contract in the light of this statistical graphic representation of the BST? A first mistake would be to conclude that the BST identifies the pathological phenomenon as a simple statistical deviation of a trait. Boorse clearly opposes this old idea.[12] What is abnormal (in statistical terms) is not necessarily pathological. The rarity of a trait is not enough to make it pathological, and this is a crucial point in the anti-psychiatry debate concerning the medicalization of deviant behaviours. The value of Boorse's approach is that it focuses on the statistical distribution of the *effectiveness of a function*, not on the *presence of a trait*. This is an important distinction, since the bell curve of the effectiveness of a function takes into account the synchronic representation (statistical approach) of a diachronic process, i.e. the whole evolutionary past of the part-function. Since "some of the past affects what is species-typical" (Boorse 1997: 66), the pathological phenomenon takes root in some biological disadvantage through a decrease in individual reproduction and survival within a

---

[11] The author has recently acknowledged that the axes of his original picture in 1997 were mislabelled. Functional efficiency should be the x-axis, and frequency the y-axis. In any case, the picture shows a bell-curve figure that measures efficiency of a part-function in a population. For a detailed analysis of this figure, see for example Ananth (2008: 117) who comments on this figure with the example of thyroid functioning.

[12] Cf. Boorse (1977: 546): statistical normality is neither a necessary nor sufficient condition of health. For an in-depth discussion of this aspect, see Ananth (2008: 18 and following pages).

population when the function concerned is no longer effective. Thus, similarly to Adolphe Quételet who referred to "constant causes", Boorse uses the role of natural selection in the evolutionary history of organisms to explain the *normality of the frequent*, albeit in a more subtle way than Quételet – who was only able to explain the normality of the frequent with reference to a divine plan.

A second mistaken interpretation of the bio-statistical theory, which is somewhat the reverse of the first, would consist in neglecting the issue of statistically typical functional effectiveness by focusing only on *biological disadvantage*.[13] This second mistake is recurrent in psychiatry, as some authors have attempted to correlate the presence of certain clinical syndromes with criteria like fertility or life expectancy in order to provide empirical evidence of their pathological character (cf. Kendell 1975).

So, by bringing together a biological and a statistical approach, Boorse provides a convincing reply to two detrimental misunderstandings in medicine in general, and in psychiatry in particular. It is however at the cost of a "statistical idealization" which has absolutely no pragmatic value. First, there is no reason why the empirical measure of any biological fact or mental functioning within a given population should reproduce the exact shape of a Gauss curve. The bell curve meant to represent the effectiveness of a particular function within a population is condemned to be an ideal diagram. Furthermore, nothing justifies the arbitrary threshold of 2.5 %[14] for a phenomenon to be qualified as pathological. Boorse is aware of the arbitrary nature of this threshold, and for once he leaves it up to clinical medicine to decide on it. But this number, which seeks to confer a biological meaning to statistical normal variation, cannot provide a consensual rationale in medicine for delimitating what should be considered as a pathological condition.[15] Lastly, and this exacerbates the previous shortcomings, very often measures of the effectiveness of a function are taken imperfectly or indirectly, as Boorse himself recognizes (Boorse 1997: 50), through the measure of biological indicators (or psychometric measures in psychopathology).

---

[13] Boorse accuses Wakefield of siding him with advocates of biological disadvantage, alongside Scadding or Kendell. Cf. Boorse (1997: 65).

[14] 95 % of what we call the normal variation of a Gaussian distribution sits within an interval that includes two typical deviations on both sides of the average. As Boorse is only interested in the lower fringe of the effectiveness curve, he comes to set the pathological phenomenon threshold around 2.5 %.

[15] To take a very simple example, the prevalence of visual impairment in France is estimated at 6.8 % for 70-year-olds, and at 14 % for 80-year-olds (Ministère des Affaires Sociales et de la Solidarité Nationale 1990). Following Boorse's model, we could consider that within each age group only the 2.5 % most visually impaired are really pathological, leaving the others to convenience medicine. Yet when we know how severe this impairment is (a person is considered to be visually impaired if the visual acuteness of their best eye, after optical correction, is lower than 4/10), it may seem strange to still be talking of a "convenience medicine". Note that Wakefield's essentialist approach, which I will discuss later on, would reach a simple and more consensual conclusion regarding this specific example that would however not necessarily be based on a more valid reasoning. His idea would be that the eyes' natural function, even for the elderly, is to see, and that this consideration alone is enough to consider the problem of visual impairment as a pathological one, regardless of age.

Together these three limitations make the bio-statistical criterion of demarcation between the normal and the pathological impossible to determine objectively, even approximately, and especially when it comes to controversial issues in psychiatry. In other words, the statistical graphic representation that Boorse gives in his 1997 paper is purely illustrative: it theoretically illustrates the fact that we can speak of a pathological phenomenon once the efficacy associated to a function is so weak that it no longer accounts for the fitness of a population. The sole value of this curve is that it reconciles the idea of *natural variation* with that of the *objectivity of the pathological phenomenon*. But it does not provide any means to quantitatively distinguish normal phenomena from pathological ones for a given function. Hence it would be wrong to think that it has any heuristic value, or that this graphic representation could lead, for example, to certain empirical research areas in medical epidemiology.

I believe it is no coincidence that no in-depth discussion exists of the potential practical application of Boorse's bio-statistical approach to the mental domain.[16] It is indeed difficult to imagine how such a theoretical definition could be productive in the field of mental diseases, even for those mental diseases that boast minimal consensus, like schizophrenia or depression. The reason is this: in the field of mental disease, the issue is not only separating the relative weights of biological and cultural factors in empirical measures; it is also being able to clearly identify the *mental functions* that are allegedly failing, and being capable of providing a scientific measure of their effectiveness within a population. Supposing for example that depression is a pathological dysfunction of sadness (see Horwitz and Wakefield 2007), the idea of one day being able to obtain a non-controversial measure of the *natural* effectiveness of sadness is simply unrealistic. In other words, the BST gives absolutely no clue, even from a theoretical point of view, how to objectively distinguish an intense period of sadness from a true depression episode.

I therefore conclude that while Boorse's work offers a thought-provoking theory to take on – albeit indirectly – the challenge set by Canguilhem of translating the difference between normal and pathological phenomena into quantitative terms, his solution stands no chance of providing clarification to any psychiatric debate. Needless to say, Boorse was aware of the pragmatic limitations of his theory, and intended to leave it up to medicine to define and classify individual disease entities using "whatever is the most scientifically convenient basis" (Boorse 1997: 64). But it seems that the bio-statistical approach he advocates is bound to flounder on difficulties in the field of psychiatry that are not simply temporary. Boorse seems to dismiss the possibility for clinical psychiatry to scientifically determine the limits of pathology, but he seems to forget that psychiatry still only has clinical medicine to fall back on to legitimate its practice. The difficulty with the BST is that it is not a "value-free account"; it is rather a purely idealized account of disease, let's say a "free-disease account" that provides no ground for discussing about real complex

---

[16] Rachel Cooper's (2005) and Jonathan Tsou's (2008), focused on the psychiatric question and carefully discuss Boorse's proposition, but both works essentially stick to the general biological objections.

diseases or real medical controversies. It would be similar to the situation where a philosopher would give a detailed definition of a "star" in astronomy, but his definition would be of no help in determining if "Venus" or "Alpha Centauri" are real stars.

Wakefield's Essentialist Approach

The Harmful Dysfunction Analysis

Jerome Wakefield, contrary to Christopher Boorse, is a specialist in the psychiatric field. His seminal article "The Concept of Mental Disorder: On The Boundary Between Biological Facts and Social Values", published in the journal *American Psychologist* in 1992, has been one of the most cited and commented upon articles in philosophy of psychiatry over the last 20 years. Unlike Boorse and his "unrepentant naturalism" (Boorse 1997: 5), Wakefield defends a weak normativist position. This amounts to recognizing the fundamental interference of social values in the psychiatric domain, but to counterbalancing this problematic aspect by highlighting a factual dysfunctional component that legitimizes psychiatry as fully belonging to the medical model. Actually, Wakefield's theoretical position is very close to that of Boorse. The difference is mainly that Wakefield is searching for a more practical tool that can be applied to the psychiatric domain, to serve as a test or filter to determine the conceptual validity (or not) of the diagnostic criteria presented in reference classifications, whereas Boorse was more concerned with promoting an objectivist position that was criticized at his time of writing. Thus, when Wakefield adds a "harm" clause to his analysis, it is more to promote the idea, in theory possible (and which Boorse ultimately agrees with), that there can be dysfunctions that do not deserve to be qualified as pathological from a clinical point of view, for the reason that they are not harmful and can even sometimes benefit an individual – Wakefield takes the speculative example of an aging mechanism dysfunction that would lengthen life (Wakefield 1992b: 384). In other words, Wakefield seems to follow in Boorse's footsteps but he steers the conceptual analysis programme towards an explicitly practical and clinical level, where the harmful component is central. In fact, in his 1997 Rebuttal, Boorse did not hesitate to present Wakefield's HDA as more of a continuation of the BST approach than a rival theory: "His own view, as I said, is essentially the BST with Wright's function theory replacing mine and a harm clause added to make dysfunction clinically important" (Boorse 1997: 66).

The similarities between the two theories are undeniable. Yet the differences are greater than they may seem. Apart from the harmful clause added as a necessary component to the definition of a pathological condition, Wakefield fulfils the condition (i) of the objectivist contract in the same manner as Boorse: the normal can be accepted as a fact, provided it is discussed within the framework of normal biological functioning. But his solution to condition (ii) is quite different. Contrary to Boorse, Wakefield seems to have refused the challenge set by Canguilhem of translating the difference between the normal and the pathological into quantitative terms. Wary of any statistical characterization of the pathological phenomenon, Wakefield explicitly adopts an *essentialist* position, whose theoretical justification draws on the theory of evolution.

What is Wakefield's criterion of distinction between the normal and the pathological based on? We could say that it is based on a difference of *essence* or *nature* which, if we carefully take the assumptions of the HDA into account, cannot be unequivocally translated by any statistical means, or even be quantitatively formalized in theoretical terms. The difference between a depression and a profound sadness, for instance, is qualitative rather than quantitative. Though it is objective and biologically rooted, this difference can only be understood in contextual terms and therefore, according to Wakefield, it requires a certain clinical tact. For the depressive person, sadness is dysfunctional, in the sense that it is obvious from a clinical perspective that it does not fulfil the biological function it is meant to fulfil in a normal individual. On the other hand, for a person who has just lost a loved one, sadness is normal (even though it may be intense): it responds to an event of loss which, no matter how vague our understanding of the biological mechanism involved may be, can be seen as the trigger for the type of normal response for which the emotion of sadness was naturally shaped by evolution. I will not discuss here the many issues that this approach raises in the specific case of depression. My objection is more general in scope and is concerned with Wakefield asserting his theory's capacity to resolve most of the nosological controversies by drawing on a certain type of essentialist reasoning. The focus of my criticism is the particular theoretical conception of biological function on which Wakefield bases his definition of the normal and the pathological. By seeking to turn a bold theoretical definition (bolder than he cares to admit) into a tool that could be used to address clinical practice, Wakefield's HDA is more vulnerable to criticism than Boorse's theory. For his definition to be useful to psychiatry, Wakefield indeed has to neglect certain theoretical subtleties that Boorse was careful to describe. To begin with, Boorse insists on the gap that separates pathology as a theoretical science from clinical medicine as a medical practice,[17] thereby setting out a conciliatory "multi-level" space (Boorse 1997: 101). Wakefield, on the other hand, offers an analysis that claims to reconcile the most traditional clinical intuitions with the most recent evolutionist hypotheses. While Boorse displays a minimum of caution with regard to lay usages of the concept of disease, Wakefield does not hesitate to appeal to common sense at different stages of the conceptual justification of his approach. Like Boorse, Wakefield fuels his analysis with common sense intuitions, but he goes much further on a certain theoretical level since he explicitly relies on an evolutionary conception of biological functions. This difference, I argue, highlights a specific frailty with Wakefield's theory that does not exist in Boorse's theory. Boorse is careful to point out that his bio-statistical theory is not based on any definitive approach

---

[17] *Ibid.*: 11: "It aims at a pathologist's concept of disease, not a clinician's […]."

to the biological function: so his work on the concept of health can be separated from any specific account of the notion of biological function (cf. Boorse 1997: 10).[18]

Wakefield's position is however far more tenuous. Although he also insists on the fact that his analysis can be separated from the etiological conception of function that he advocates, it is important to point out that *all the practical and clinical value he attributes to the HDA is nevertheless founded upon it*. Remove the evolutionary anchoring of Wakefield's HDA, and it loses all its flavour. But there is another problem that appears with the evolutionary anchoring of his position: Wakefield must justify the *unity* of the theoretical and lay concept of function. Indeed, the idea that a function is a selected effect (according to etiological theory) is not as intuitive as it may appear for the clinician. And as Kenneth Schaffner (1993) has shown, it is far from coinciding with the idea of "function" in use in the biomedical sciences.[19] Wakefield has first to resolve this apparent hiatus.

"Black Box Essentialism"

Even though Wakefield has given "black box essentialism" a predominant place in his work for a long time,[20] he only started to refer to it later in response to certain objections raised against the HDA. I think that this conception, which is presented as applicable to all scientific concepts, is both the core and the weakness of Wakefield's analysis of mental disorder.

Clearly, Wakefield first started to refer to this conception at a time when he needed to meet two expectations to defend his HDA: (a) he had to ensure the unity of the concept of function, whether in the psychiatric field, the medical field or the biological field in general; and (b) he had to propose a method that would make it possible to define a set of mental dysfunctions, and in a way that could seem legitimate in spite of the persistent tenuousness of scientific knowledge in psychiatry.

---

[18] Indeed, one could criticize Boorse for having a tendency to defend his bio-statistical theory, through his articles, by readily drawing on the conception of function that he always advocated and which is goal oriented. We should note however that the concept of function which he promotes as a "current contribution to a goal" is sufficiently flexible and broad to encompass most of the particular meanings of the term "function", as much in biology as in medical practice, all the while being in agreement with common usage.

[19] Schaffner stresses that evolutionary functional explanations are needed in the biomedical sciences, yet are "so empirically fragile that we could even call them metaphysical" (Schaffner 1993: 389–390).

[20] Wakefield has been referring to "black box essentialism" for a long time in his work (a reference can be found as early as 1991 in an article dealing with the question of emotions in Freud's work). As he himself recognizes, this theory is largely inspired by Putnam's account of the meaning of theoretical terms. With regard to the specific problem I am interested in here, however, the first noteworthy reference dates back to 1997.

The Unity of the Concept of Function

Wakefield's HDA is grounded in the methodological framework of conceptual analysis which he sums up as follows:

> In a conceptual analysis, proposed accounts of a concept are tested against relatively uncontroversial and widely shared judgments about what does and does not fall under the concept. To the degree that the analysis explains these uncontroversial judgments, it is considered confirmed, and a sufficiently confirmed analysis may then be used as a guide in thinking about more controversial cases. (Wakefield 1992a: 233)

While he notes that consensual judgements are not necessarily assumed to be correct, a good analysis must nevertheless allow for discerning correct judgements from those that are not. The aim of using the method of conceptual analysis is to provide a *descriptive definition* of mental disorder, which could adequately be used by physicians on an everyday basis, and even in common understanding. Any discrepancy with the ordinary meaning presents a significant problem for the definition, which must be accounted for.[21] The explicit goal of the HDA is to *describe* and not to *reform* our conception of the pathological phenomenon.

Yet in his 1992 article Wakefield was already promoting his approach with two assumptions: first, that only the *evolutionist* concept of function can potentially provide a scientific basis for identifying mental dysfunctions; but second, that there is nevertheless a profound unity in the concept of function, as it is intuitively understood and used in the medical sciences, and as it is defined from the perspective of evolutionary theory. In a discussion with Christopher Megone about the meaning of the notion of function in Aristotle's work (Wakefield 2000), Wakefield elaborates on his position regarding an important issue raised by Ruth Millikan and Karen Neander, i.e. the difficulty of reconciling the assumptions underpinning conceptual analysis with an etiological conception of function. By putting a particular conception of natural function in biology at the heart of his conceptual analysis of the pathological phenomenon, Wakefield has to take part in the highly discussed debate in philosophy of biology opened by Wright's and Cummins' papers in the 1970s. Namely he has to respond to a very strong criticism from the American philosopher Ruth Millikan who sees a discrepancy, for certain terms such as "function", between common sense characterizations and theoretical characterizations. Millikan, who endorses the *reformist character* of her theoretical conception of biological functions, expresses her grievances against the conceptual analysis method as it is used in philosophy, which is too often drawn upon to justify in the analytical tradition the well-founded ground of our common concepts:

> Now I firmly believe that "conceptual analysis", taken as a search for necessary and sufficient conditions for the application of terms, or as a search for criteria for application by reference to which a term has the *meaning* it has, is a confused program, a philosophical chimera, a squaring of the circle, the misconceived child of a mistaken view of the nature of language and thought. (Millikan 1989: 290)

---

[21] For a more detailed discussion about the limits of conceptual analysis concerning this issue, see Aucouturier and Demazeux (2012).

Regarding this specific issue, the philosopher Karen Neander answered Millikan by defending the benefits of conceptual analysis and by showing that it was compatible and even complementary with the search for a *theoretical definition* of certain scientific terms (Neander 1991). Wakefield, whose entire approach is based on conceptual analysis, criticizes both authors by arguing that the concept of function is not *directly* linked to the idea of natural selection, neither through a conceptual analysis, nor through any theoretical definition. "It is a scientific discovery, not a conceptual truth, that functions exist because of natural selection", he claims (Wakefield 2000: 39). He justifies this with reference to a two-stage process: there would first be a common meaning of the concept of function, intuitively shared by all, throughout time (and therefore as much by Aristotle, by Harvey or by ourselves). This would be the common meaning highlighted by the conceptual analysis. Then there would be a "modern discovery", which would show that natural selection explains the "essence" of functional processes. To ensure the continuity between the two stages, Wakefield refers to what he called "black box essentialism", a concept he advocates following the works of Putnam, Kripke and Searle on *natural kinds*. Here is the definition he proposes:

> It is "essentialist" because the criterion for membership consists not of the properties that originally inspired one to define the kind, but of some underlying "essential" property that explains the observed surface features. It is "black box" essentialism because, rather than claiming that scientific concepts are defined by specific essences (e.g., "water is H_2O"), this view asserts that such definitions remain agnostic on the identity of the essence and leave its discovery to science. (*Ibid.*: 36)

This reference to "black box essentialism" is crucial and I shall return to it. But let us first consider what, for psychiatry, comprises this very theoretical issue surrounding the unity of the concept of function. Tacitly, the issue is to defend *the continuity and perennial appeal of the psychiatric tradition*. Aristotle shared the same concept of function as us, even though he may not have known of the modern explanation provided by evolutionary theory. Likewise, eighteenth century alienists shared the same idea of mental disease as us, even though they were incapable of justifying it scientifically. Wakefield refuses to see any discontinuity in the history of psychiatry; for him, the "Darwinian psychiatry"[22] that he seeks somehow to promote, despite being in its early days, does not depart from traditional psychiatric discourse. The HDA does not impose any fundamental break with or reform of psychiatric discourse: rather, it seeks to consolidate it.

---

[22] This label has been popularized by McGuire and Troisi following Nesse and Williams. Wakefield actually uses it very little, and does not accept its theoretical premises. The reason is simple yet very significant: as Murphy (2005) highlighted, the Darwinian approach to mental disorders integrates three possible types of explanation: internal dysfunction (breakdown); the fact, for an organism, of no longer being adapted to an environment that has changed rapidly (mismatch); and the paradoxical possibility of a mental disorder being an adaptive advantage (persistence). Yet Wakefield's HDA is only compatible with the first type of explanation (breakdown), which puts it at odds with many evolutionary hypotheses.

The Heuristic Value of HDA

Relying on the "black box essentialism" theory, Wakefield attributes some heuristic virtues to his approach. But the difficulty for him is the following: without a method allowing for legitimately defining the domain of mental disorders in the absence of ascertained scientific knowledge, his analysis would be of no use in psychiatry. What would be the point of the HDA, if nothing, in our current state of knowledge, could legitimately allow us to infer the existence of certain harmful dysfunctions? At best, the definition could serve as a "regulatory ideal", to quote Kant. But it would certainly not be of any help in resolving the current issues in psychiatry. Yet Wakefield's *coup de force* lies in him drawing on an evolutionary approach to legitimate most (if not all) of the categories of the American official classification in psychiatry, the *Diagnostic and Statistical Manual of Mental Disorders* (DSM):

> If one looks down the list of disorders in the DSM, it is apparent that by and large it is a list of the various ways that something can go wrong with the seemingly designed features of the mind. Very roughly, psychotic disorders involve failures of thought processes to work as designed; anxiety disorders involve failures of anxiety- and fear-generating mechanisms to work as designed; depressive disorders involve failures of sadness and loss-response regulating mechanisms; disruptive behavior disorders of children involve failures of socialization processes and processes underlying conscience and social cooperation; sleep disorders involve failure of sleep processes to function properly; sexual dysfunctions involve failures of various mechanisms involved in sexual motivation and response; eating disorders involve failures of appetitive mechanisms, and so on. There is a certain amount of nonsense in the DSM and criteria are often overly inclusive. However, the vast majority of categories are inspired by conditions that even a lay person would correctly recognize as a failure of designed functioning. (Wakefield 2007: 152)

This long citation is very illuminating, since it encapsulates the ambivalence of Wakefield's position. On the one hand, the author insists on the highly evident reality of the disorders he lists, and does not hesitate to assert that "even a lay person would fully recognize" the apparent validity of the categories of the DSM-IV. Wakefield has always recognized this "evident reality", furthermore considering that the term "dysfunction" finds its roots in the vulgar intuition that "something has gone wrong 'in the person'" (Wakefield 1992a: 240). But on the other hand, he has insisted on several occasions in his work on the fact that the inference in question is only "hypothetical", that it is "risky"[23] in the sense that it can be falsified, and furthermore on the fact that it must be steered cautiously according to all the circumstantial evidence[24] at hand. Note that the issue here, for psychiatry, is to defend

---

[23] Wakefield (1999a: 376): "I argue here that failure of a naturally selected function is necessary for disorder. This is a highly risky claim: it can be falsified by just one clear example of a disorder that is not an evolutionary dysfunction." Or Wakefield (1999b: 967): "The attribution of disorder ultimately involves a broad theoretical hypothesis that the cause of the symptoms involves a dysfunction, and this hypothesis can be falsified."

[24] Wakefield (2003: 971): "Obviously, in our present state of ignorance, we judge that there is a failure of some internal mechanism from circumstantial evidence that makes such a hypothesis overall the most plausible, just as, long before we understood etiology, we correctly recognized blindness and paralysis as disorders based on the circumstantial evidence of an individual's inability

the validity of its current categories after having consecrated the unity of its tradition. Even though in the current state of affairs, the classification of mental disorders looks more like a "troubleshooting guide"[25] than a scientific classification, the "black box essentialism" approach seems to allow for the type of inference currently made by common sense and clinicians, by targeting an essential dysfunction underlying clinical manifestations even when the essence of this dysfunction remains undetermined. Wakefield goes so far as to vaunt the salutary ecumenism of this approach as, he claims, the indeterminacy of the black box makes it possible to transcend the theoretical divisions that undermine psychiatry. But are this ecumenism and this theoretical indeterminacy epistemologically satisfying?

The Troubled Waters of Essentialist Inference...

Every time Wakefield refers to the "black box essentialism" approach, he draws on the paradigmatic example of water. This occurs in at least six instances in six different articles: in his criticism of Hans Eysenck's too restrictive essentialism (Wakefield 1997); in a defence of the general validity of the HDA (Wakefield 1999c); in the discussion I mentioned earlier about the concept of function in response to Millikan and Neander (Wakefield 2000); in a critical reflexion on Michel Foucault's ideas (Wakefield 2002); in an argument about the compatibility of his position with that of Ian Hacking (Wakefield 2003); and lastly, in a critical appraisal of Paul Meehl's "open concepts" (Wakefield 2004).

Each time the issue is very different: in one case he is seeking to justify the unity through history of the concept of function; in another he highlights the causal relationship between an underlying dysfunction and the surface symptomatology of a disorder; against Hacking, he insists quite to the contrary on the difference between a fixed essence of a disorder and the plasticity of its clinical manifestations; against Foucault's constructivist trend, Wakefield defends the natural tendency for the human mind to make essentialist inferences; and lastly, against Paul Meehl, Wakefield refutes the idea that psychological concepts are always implicitly defined by theories.

In all cases, Wakefield refers to the same paradigmatic example of water, in other words of a natural substance whose surface properties can be identified (any common dictionary will sum them up quite well by presenting water as a "colorless, odourless and transparent liquid that is insipid when pure"), and whose essence (or "inner nature") science was only able to characterize late in time (praise be to Lavoisier and the chemical formula H_2O). The analogy with mental disorders is clear: according that the surface properties are the clinical symptoms of a mental disorder, it is logical to infer the existence of a dysfunction that provides an in-depth

---

to see or move under conditions in which it was presumed that normally functioning individuals would be able to see or move, and where alternative explanations seemed unlikely."

[25] Wakefield (1999b: 971): "[...] The DSM's logical structure is closer to that of a trouble-shooting guide than to that of a theory or research program."

explanation of the reliability and validity of the clinical pattern, even when science does not yet know how to definitively identify this dysfunction. The example of water provides the paradigm of such a kind of legitimate inference:

> Even the simplest scientific notions, such as that there is a substance water that is distinguishable from other substances by some internal structural property and that liquid water, ice and steam are all instances of this substance [...], depend on inferring the existence of underlying causal structures that explain appearances long before the underlying structure can be identified. (Wakefield 1999b: 987)

Wakefield takes the analogy between the concept of water and the concept of mental disorder very seriously. The only difference Wakefield acknowledges between these two concepts is when he criticizes Eysenck's behavioural essentialism. He then argues that the concept of mental disorder, contrary to that of water, is not *purely* an essentialist concept: it is, as he calls it, a "cause-effects" concept, in the sense that its definition *also* depends on the surface properties, since only a dysfunction with harmful effects in the clinical sense can be a mental disorder. In the same way that one is only a criminal if they have committed a crime, so a dysfunction, by virtue of the HDA, is only a mental disorder if it is harmful according to some social norms.[26] But for the rest, the analogy not only serves as an illustration: in both cases, the inferential process is the same. It follows the same logic of scientific discovery, it tolerates the same circumstantial variations in surface traits,[27] and it is universally shared by lay people.[28]

Yet the importance that Wakefield gives to this simple analogy points to the weakness of his theory. For that should be the starting point: is this analogy between the concept of water and the concept of mental disorder even legitimate? Between the application of "black box essentialism" to a natural element and its application to a set of conditions which, throughout human history, never presented itself as clearly as Wakefield often claims, there is a difference that calls for a certain caution. Yet Wakefield never really considers the possibility that much of the mental disorders currently identified could be anything other than natural kinds. His theory certainly *presupposes* that a mental disorder is something like a natural kind. Contrary to Paul Meehl, for instance, who saw "open concepts" as one reason for explaining the lack of progress in psychopathology, Wakefield considers that even if people are wrong about their particular theories concerning mental disorders, they are right to attribute a fixed meaning to most of the recognized mental disorders.

---

[26] Thus there would be good grounds, according to Wakefield, to continue calling a substance with the chemical property H_2O "water" even if it does not present the surface property normally associated with it (for example, when water in a still unknown state is discovered on another planet). For a more detailed discussion concerning this argument, see LaPorte 2004: chap. 4 and Hendry 2010.

[27] Just as water and ice, Wakefield writes, share a same substance with varying surface properties depending on circumstances (like temperature), so too the clinical manifestations of a disorder can vary according to (social or cultural) circumstances without calling into question the unity or reality of the disorder.

[28] This is the core of the argument made in the 2002 *"Fixing a Foucault Sandwich"*.

In other words, even if official criteria for some mental disorder can be strongly dependant of the social context, Wakefield assumes that "criteria and theories are open, but concepts have fixed meanings" (Wakefield 2004: 78).

Unfortunately, neither the fact that human beings, naturally, are prone to making essentialist inferences on all sorts of things, nor the fact more specifically that they spontaneously tend to think of a set of negative mental states as pointing to the idea that "something is wrong with a person" are sufficient guarantees for thinking that mental disorders are what human beings believe they are. Gaston Bachelard, in *The Formation of the Scientific Mind*, also agreed that human beings have a tendency to essentialize. But from his thorough investigation in the history of science, he drew a conclusion diametrically opposed to Wakefield's: the essentialist way of thinking often constitutes more of an obstacle to scientific progress than a guarantee of its advancement. From this point of view, Wakefield does seem to stumble at "the substantialist obstacle" identified by Bachelard – as recurrent in the history of chemistry as in the history of medicine – which consists in believing that "*substance has an inside*, or better, that substance is an inside" (Bachelard 2002: 106), and in spontaneously trusting "the light of that intuition which puts us at the *heart* of reality" (Ibid.: 108). Indeed, not only does the opposition between "surface properties" and "underlying essence" underpin Wakefield's thinking without any epistemological or ontological clarification ever being provided, but he also maintains that the natural tendency to essentialize is an "innate cognitive tendency" which should in a way just be accepted: "In any event, if we do tend to be essentialists by nature, then we had better confront that fact" (Wakefield 2002: 27). The problem, Wakefield mentions in passing, is that "of course, essentialism can mislead us" (*Ibid.*: 26).

Yet that essentialism can lead us astray is the fundamental issue. For what guarantees us that surface traits always converge towards an underlying essence? From a scientific perspective, what legitimates the value of our essentialist inferences? By focusing on the ideal example of water, is there not a risk of neglecting far more delicate and controversial examples that are however closer to the type of danger that psychiatry wishes to stave off, for instance the hazardous inferences that have led some to deduce from a set of superficial properties (size and shape of the skull, physiognomy, etc.) a common nature for murderers or intellectual inferiority as a characteristic of certain races? Why is such a risk not seriously taken into account? As long as science does not provide any confirmation, the essentialist inference remains a hypothesis, and its comparison with other fruitful inferences from the past is certainly not enough to guarantee its value. As we have seen, on this specific issue Wakefield is highly ambivalent. He constantly switches between a weak version of his theory, which consists in recognizing the usefulness and even necessity of psychiatry having to make do with inferences that are only hypothetical, and a strong version that advocates definitively consecrating the validity of the inferences informing psychiatric labels, which he sees no need to call into question. In a 2006 article, Wakefield nevertheless seemed aware of the difficulty he faced:

> It clearly shows from historical and anthropological accounts that values, norms and ideologies profoundly influence what people believe to be natural functions, particularly when there is

a lack of scientific understanding of what is functional or dysfunctional (as is currently the case for many mental aspects). (Wakefield 2006: 43)

Either Wakefield thinks that there is currently a lack of scientific understanding of what is functional or dysfunctional in psychiatry, or he considers that the evolutionary approach is now capable of filling the void. In other words, either the HDA provides nothing more than the ancient and fragile conviction, prone to all sorts of excesses, that "something is wrong with the person", or it claims to provide a sounder confirmation by drawing on the contributions of evolutionary psychology. What is Wakefield's position faced with this dilemma? From a general point of view, the evolutionary conception of function is, as I have said, clearly presented as this *scientific discovery* that now makes it possible to retrospectively legitimate psychiatry's founding intuitive inference concerning the nature of the concept of mental disorder, as chemistry once did for water. This conclusion seems a little bit hasty, but what is more interesting here is how Wakefield plans to differentiate normal phenomena from pathological ones. Considering psychiatry's specific diagnostic labels, Wakefield recognizes that each particular label is only hypothetical, at best resting on all the circumstantial evidence available: "the HD analysis is an analysis of the concept of disorder, not a theory of the mechanisms or dysfunctions underlying disorders" (Wakefield 2003: 978). And yet "the HD analysis offers a framework for constructing" particular theories (*Ibid.*) for each specific mental disorder. To date, the example Wakefield has used and elaborated upon the most is depression, which he accounts for based on cross-cultural, developmental, comparative, and other evidence (e.g. grief, attachment, facial expression, etc.) (Wakefield 2003: 978; see also Horwitz and Wakefield 2007), a dysfunction of sadness understood as a natural mechanism for managing responses to loss phenomena. In the specific case of depression, Wakefield has detailed the observations and scientific evidence that make, according to him, his hypothesis plausible. But this does not prevent him, in other places and for other disorders, from reaching conclusions with surprising assurance:

However, most DSM categories represent failures in functions for which the species-typical designed nature is not seriously in dispute, such as sexual arousal, sleep, fear, sadness, thought, motivation and so on.[29] (Wakefield 1999b: 986)

Evidence used to justify psychiatric labels by referring to faculties of the mind as vague and encompassing as "thought" and "motivation" is very weak and embarrassing, to say the least. The serious problem is that Wakefield does not seem to take stock of the criticism levelled at him that the type of inference authorized by the HDA, due to its indeterminate nature, makes it very easy to build *ad hoc theories* to justify the apparent validity of any random diagnostic label. Wakefield himself is not immune to this risk; a few examples taken from his articles will suffice to

---

[29] This quote should be contrasted with the following: "Whether dysfunction actually exists is an empirical issue not addressed by the HD analysis, which must be assessed case by case. If there is no reason to infer dysfunction, then the HD analysis disallows disorder attribution, contra DSM" (Wakefield 2003: 971).

show that it is very real. The first example is drawn from his discussion on a label introduced into the DSM-III-R that generated intense controversy: Oppositional Defiant Disorder. Wakefield made the following comment:

> Instead of DSM-III-R's clearly invalid statistical and operational definition of oppositional-defiant disorder, which labels children as disordered on the basis of greater-than-average negative behavior, one may conclude that attribution of oppositional-defiant disorder is based on an inference from behavior to a dysfunction in certain aggression-inhibitory mechanisms (vague identifying descriptions are acceptable when more specific knowledge is lacking). (Wakefield 1993: 170–171)

To put it mildly, we could say that Wakefield was only trying here to "save the phenomena" (namely the clinical description of the disorder), by reorienting the description and research towards the only hypothesis which he deemed legitimate in relation to the HDA. But it would be difficult to see in the redefinition he advocates anything more than a rhetorical pirouette. Worse, one can even wonder if in doing so, Wakefield is not making the label in question immune to any possible criticism by making it hard to falsify: what current empirical evidence could call into question such a hypothesis? And what golden age of science are we meant to wait for to be able to clearly settle the matter? The second example, taken from the same 1993 article, concerns the characterization of hypochondria that Wakefield casually mentions in passing in his discussion:

> The rational use of information is a cardinal natural function of the mind. If rational mechanisms are not functioning properly, then disorder may be inferred even when ignorance is a causal factor. For example, some hypochondriacal patients feel irrationally distressed about certain disease possibilities until they get conclusive laboratory evidence that they do not have the disease. The lack of such conclusive information is, then, a necessary causal condition for their distress, but that does not imply that their condition is not a disorder because the relation between the information that is lacking and the resulting anxiety is irrational and implies that the cognitive and affective systems are not functioning properly. (Wakefield 1993: 168)

Not only does Wakefield blithely redefine the theory and clinical symptoms of hypochondria,[30] he does so based on a hypothesis that permits all sorts of excesses. In this sense, any strong and disproportionate anxiety will as such be pathological, and will be associated with a physiological if not mental dysfunction. It may even be that any irrational reasoning that jeopardizes this "natural cardinal function of the mind", which allows for rationally processing information, will have to be considered as pathological. In any case, justifying the possibility of hypochondria being a mental pathology based on the ever so vague hypothesis of a dysfunction of "the cognitive and affective systems" is indeed not very different from simply justifying it with the conviction that "something 'in the person' is not functioning as it should be".

---

[30] Hypochondria is traditionally associated with neurosis, hence with a type of condition that has little to do with rational mechanisms of information processing. Furthermore, clinically speaking, hypochondria is considered to occur precisely when the patient continues to be convinced they are ill, even when the tests requested provide reassurance.

The last and most illustrative example pertains to the diagnosis of orgasm disorder in women. Defending the DSM-III-R, Wakefield argues in a 1992 article that the criteria provided by the official classification, when met, "come close to demonstrating that there is indeed a dysfunction behind the lack of orgasms".[31] The DSM-III-R, according to him, manages to clearly distinguish between this alleged pathological disorder, which involves a dysfunction of a woman's "internal orgasmic mechanisms" (*Ibid.*), and the mere absence of orgasms which can be linked, as we all know, to numerous external factors. The assurance with which this very doubtful essentialist inference is presented is seriously at odds with Wakefield's emphasis on the "factual" requirement of his definition. More ironically, most current evolutionary hypotheses converge towards the idea that female orgasm is very unlikely to ever have been an effect selected during evolution (cf. Lloyd 2005), which would imply that it has never been a natural function.

But this last example also highlights a theoretical incongruity within the HDA. Let us imagine that in light of some new evidence, Wakefield recognizes his error. This would mean that what he previously considered as a mental disorder worthy of attention would cease from one day to the next to be seen as such, and would logically have to be removed from the classification of mental disorders. This therefore amounts to saying that for a patient presenting exactly the same clinical picture, with exactly the same personal and psychological suffering, what ultimately determines the legitimacy of their psychiatric treatment is the discovery that for Pleistocene women orgasm may or may not have constituted a selective advantage. It is absurd.[32]

It would be unfair to reduce all of Wakefield's work, so insightful in many respects, to these three contentious examples. I do however think that the examples point, at a theoretical level, to the shortcomings of his essentialist approach. When it does not directly rely on intuition, this approach draws on evolutionary hypotheses which are either hard to falsify, or grossly patched up when needed, or else simply of little relevance to psychiatry. In any event, factual evidence is clearly lacking in this domain. It is also doubtful whether, in terms of methodology, "black box essentialism" would actually be able to provide psychiatry with anything more than "a retreat to obscurity".[33] Far from offering clarification to the scientific status of psychiatric labels, the HDA only reinforces an intuitive medical conviction. The evolutionary approach that Wakefield advocates is much too permissive, and with or without it, his theory boils down to the following observation by Dominic Murphy and Robert Woolfolk:

---

[31] Wakefield (1992a: 244): "Despite lack of guidance from the definition of disorder, DSM-III-R criterion for inhibited female orgasm manages to rectify Masters and Johnson's (1970) error. It does so by adding a series of requirements that, if met, come close to demonstrating that there is indeed a dysfunction behind the lack of orgasms."

[32] A closer look at Wakefield's "conceptual experiments" in a 1999 text shows that what ultimately determines the functionality of a trait is evolutionary history. In this respect, Wakefield's position is close to Millikan's, even though he does not take responsibility for this. Cf. Wakefield (1999a).

[33] The expression comes from Houts. Cf. Wakefield (2003: 982).

> If Wakefield thinks of his labels as just place-holders for whatever internal basis a disorder might have, then his theory is much more modest than it appears at first. The theory becomes more plausible but much less interesting. The theory would assert only that if there is some pathology, something unspecified must have gone wrong in the mind. (Murphy and Woolfolk 2000: 249)

Everything goes to suggest that the HDA has nothing more to offer for the determination of the normal and the pathological. Yet it actually claims to shed light on at least part of the *essence* of that which dysfunctions. Here is the core of my criticism: whereas Wakefield confidently claims that functional explanations can "be plausible and very useful even when little is known about the actual nature of a mechanism" (Wakefield 1992b: 382), I would like to emphasize that, on the contrary, functional explanations can be misleading and severely detrimental to science, especially when we rely on intuitions to develop them and when very little is known about the real nature of the mechanism implied. Not so long ago the great Descartes, relying on intuitive evidence, had concluded that the heart's function was to keep the blood warm and to rarefy it. And that was not for lack of close observation, nor because he was not accustomed to "distinguishing the real reasons from the seemingly real ones."

Conclusion

Unfortunately, entrusting science with settling contentious issues is not enough to make it capable of doing so. The weakness of both Boorse's and Wakefield's objectivist approaches lies in the fact that they are "factual" approaches that remain very vague about the kind of "facts" on which they draw. I have endeavoured to show that this indeterminacy was not just due to the limitation of current psychiatric knowledge. Nor it is just due to a definitional enterprise that leaves it up to science to provide the material evidence for progress, specifying only its formal content. Indeed, Wakefield's and Boorse's accounts both remain undecided regarding the conditions necessary to objectively distinguish the normal and the pathological, especially in the psychiatric field. Boorse does show that psychiatry has good reasons to assert its status as a genuine branch of medicine. He unfortunately does not manage to provide a convincing criterion of distinction between the normal and the pathological that could clarify psychiatric controversies. Wakefield seeks to provide a more useful analysis in this respect, but at the cost of an essentialist inference which, by passing off problematical functional assignations as obvious, is far from having proven its full relevance to psychopathology.

The discussion about functions in psychiatry, which has been fuelled mainly by a debate imported from the philosophy of biology, is perhaps not bound to remain in such a state of indeterminacy. We must however remember that its current utility for psychiatry is only negative: it provides a picture of what psychiatry should not be. But it certainly does not offer it the means, in the current state of knowledge, of supporting the scientific nature of its enterprise. To expect more of it would mean risking replacing the excesses of anti-psychiatrist authors, who systematically suspect all psychiatric labels of concealing insidiously policing behaviours, with the opposite excess, which consists in overestimating the scientific weight of some common intuitions underlying the majority of psychiatric labels. The best way of dealing with this problem would probably be to simply give up the idea of setting a definite biological criterion of the normal and the pathological for all mental disorders. Ultimately, philosophical thought on psychiatry may benefit from focusing more on the complexity of always singular and regional reasons, whether clinical, scientific or socio-political, which motivate the inclusion of a given psychological suffering in the long list of mental disorders.

References

Ananth M (2008) In defense of an evolutionary concept of health: nature, norms, and human biology. Ashgate Studies in Applied Ethics, Aldershot

Aucouturier V, Demazeux S (2012) The concept of mental disorder. In: Carel H, Cooper R (eds) Health, illness and disease: philosophical essays. Durham, Acumen, pp 75–89

Bachelard G (2002) The formation of the scientific mind: a contribution to a psychoanalysis of objective knowledge. Clinamen, Manchester

Bernard C (1984) Introduction à l'étude de la médecine expérimentale. Champs Flammarion, Paris

Boorse C (1975) On the distinction between disease and illness. Philos Public Aff 5:49–68

Boorse C (1976) What a theory of mental health should be. J Theory Soc Behav 6(1):61–84

Boorse C (1977) Health as a theoretical concept. Philos Sci 44(4):542–573

Boorse C (1997) A rebuttal on health. In: Humber J, Almeder R (eds) What is disease? Humana Press, Totowa, pp 1–135

Canguilhem G (1991) The normal and the pathological. Zone Books, New York

Cooper R (2005) Classifying madness: a philosophical examination of the diagnostic and statistical manual of mental disorders. Springer, Dordrecht

Cummins R (1975) Functional analysis. J Philos 72:741–760

Davidson D (1963) Actions, reasons, and causes. J Philos 60:685–700

Fulford KWM (2001) What is (mental) disease? An open letter to Christopher Boorse. J Med Ethics 27:80–85

Giroux É (2009) Définir objectivement la santé : une évaluation du concept bio-statistique de Boorse à partir de l'épidémiologie moderne. Revue philosophique de la France et de l'étranger 134(1):35–58

Hendry RF (2010) The elements and conceptual change. In: Beebee H, Sabbarton-Leary N (eds) The semantics and metaphysics of natural kinds. Routledge, New York, pp 137–158

Horwitz AV, Wakefield JC (2007) The loss of sadness. How psychiatry transformed normal sorrow into depressive disorder. Oxford University Press, New York

Kendell RE (1975) The concept of disease and its implications for psychiatry. Br J Psychiatry 127:305–315

LaPorte J (2004) Natural kinds and conceptual change. Cambridge University Press, Cambridge

Lloyd E (2005) The case of the female orgasm: bias in the science of evolution. Harvard University Press, Cambridge

Masters WH, Johnson VE (1970) Human sexual inadequacy. Little, Brown and Company, Boston

Millikan R (1989) In defense of proper functions. Philos Sci 56(2):288–302

Ministère des Affaires Sociales et de la Solidarité Nationale (1990) Les Handicapés: Chiffres repères 1990. La documentation française, Paris

Murphy D (2005) Can evolution explain insanity? Biol Philos 20:745–766

Murphy D, Woolfolk RL (2000) The harmful dysfunction analysis of mental disorder. Philos Psychiatry Psychol 7(4):241–252

Neander K (1991) Functions as selected effects: the conceptual analysis defense. Philos Sci 58:168–184

Pickering N (2006) The metaphor of mental illness. Oxford University Press, Oxford

Schaffner KF (1993) Discovery and explanation in biology and medicine. University of Chicago Press, Chicago

Szasz T (1960) The myth of mental illness. Am Psychol 15:113–118

Tsou J (2008) The reality and classification of mental disorders: issues in the philosophy of psychiatry. PhD thesis, Chicago University, Chicago

Wakefield JC (1992a) Disorder as harmful dysfunction: a conceptual critique of DSM-III-R's definition of mental disorder. Psychol Rev 99:232–247

Wakefield JC (1992b) The concept of mental disorder: on the boundary between biological facts and social values. Am Psychol 47:373–388

Wakefield JC (1993) Limits of operationalization: a critique of Spitzer and Endicott's (1978) proposed operational criteria for mental disorder. J Abnorm Psychol 102:160–172

Wakefield JC (1997) Diagnosing DSM-IV, Part 2: Eysenck (1986) and the essentialist fallacy. Behav Res Ther 35(7):651–665

Wakefield JC (1999a) Evolutionary versus prototype analyses of the concept of disorder. J Abnorm Psychol 108:374–399

Wakefield JC (1999b) Philosophy of science and the progressiveness of the DSM's theory-neutral nosology: response to Follette and Houts, Part 1. Behav Res Ther 37:963–999

Wakefield JC (1999c) Disorder as a black box essentialist concept. J Abnorm Psychol 108:465–472

Wakefield JC (2000) Aristotle as sociobiologist: The 'function of a human being' argument, black box essentialism, and the concept of mental disorder. Philos Psychiatry Psychol 7(1):17–44

Wakefield JC (2002) Fixing a Foucault sandwich: cognitive universals and cultural particulars in the concept of mental disorder. In: Cerulo KA (ed) Culture in mind: toward a sociology of culture and cognition. Routledge, New York, pp 245–266

Wakefield JC (2003) Dysfunction as a factual component of disorder: Reply to Houts, Part 2. Behav Res Ther 41:969–990

Wakefield JC (2004) The myth of open concepts: Meehl's analysis of construct meaning versus black box essentialism. Appl Prev Psychol 11:77–82

Wakefield JC (2006) Fait et valeur dans le concept de trouble mental: le trouble en tant que dysfonction préjudiciable. Philosophiques 33(1):37–64

Wakefield JC (2007) The concept of mental disorder: diagnostic implications of the harmful dysfunction analysis. World Psychiatry 6:149–156

Wright L (1973) Function. Philos Rev 82:139–168

Emerging Disease and the Evolution of Virulence: The Case of the 1918–1919 Influenza Pandemic

Pierre-Olivier Méthot and Samuel Alizon

Abstract "Why do parasites harm their host?" is a recurrent question in evolutionary biology and ecology, and has several implications for the biomedical sciences, particularly public health and epidemiology. Contrasting the meaning(s) of the concept of "virulence" in molecular pathology and evolutionary ecology, we review different explanations proposed as to why, and under what conditions, parasites cause harm to their host: whereas the former uses molecular techniques and concepts to explain changes and the nature of virulence seen as a categorical trait, the latter conceptualizes virulence as a phenotypic quantitative trait (usually related to a reduction in the host's fitness). After describing the biology of emerging influenza viruses we illustrate how the ecological and the molecular approaches provide distinct (but incomplete) explanations of the 1918–19 influenza pandemic. We suggest that an evolutionary approach is necessary to understand the dynamics of disease transmission but that a broader understanding of virulence will ultimately benefit from articulating and integrating the ecological dynamics with cellular mechanisms of virulence. Both ecological and functional perspectives on host-pathogens' interactions are required to answer the opening question but also to devise appropriate health-care measures in order to prevent (and predict?) future influenza pandemics and other emerging threats. Finally, the difficult co-existence of distinct explanatory frameworks reflects the fact that scientists can work on a same problem using various methodologies but it also highlights the enduring tension between two scientific styles of practice in biomedicine.

P.-O. Méthot (✉)
Faculté de philosophie, Pavillon Félix-Antoine Savard, Université Laval,
2325 rue des bibliothèques, Québec (QC), G1V 0A6, Canada

Centre interuniversitaire de recherche sur la science et la technologie, Université du Québec
à Montréal, C.P. 8888, succ. Centre-ville, Montréal (Qc), H3C 3P8, Canada
e-mail: p.olivier.methot@gmail.com

S. Alizon
Laboratoire MIVEGEC (UMR CNRS 5290, IRD 224, UM1, UM2), Montpellier, France

P. Huneman et al. (eds.), *Classification, Disease and Evidence*, History,
Philosophy and Theory of the Life Sciences 7, DOI 10.1007/978-94-017-8887-8_5,
© Springer Science+Business Media Dordrecht 2015

Introduction

The question "why do parasites harm their hosts?" is recurrent in evolutionary biology and ecology, and has several implications for the medical sciences, particularly public health and epidemiology.[1] The question is perplexing because of its paradoxical aspect. Indeed, one wonders why natural selection favours high virulence if this inevitably results in both the host and the pathogen's deaths. Shouldn't host and pathogens[2] peacefully coevolve, and thus maximize both their chances of survival, instead of engaging in a near-infinite arms race? Very much along these lines, *The Lives of a Cell* (1974) by American physician Lewis Thomas reflected the conviction that "there is nothing to be gained, in an evolutionary sense, by the capacity to cause illness or death" (Thomas 1974, 77). Thomas' views on the nature of disease were once widely accepted among medical scientists during the past century.[3] The possibility of eradicating diseases like smallpox, combined with the belief that evolution was going to naturally wipe out infections, worked together in supporting the idea of the end of infectious diseases (Levins 1994). Physician and epidemiologist Aidan Cockburn, for instance, stated confidently: "it seems reasonable to anticipate that within some measurable time, such as 100 years, all the major infections will have disappeared" (Cockburn 1963, 150). Following the improved control over infections provided by vaccines, antibiotics, and chemotherapy, biomedical authorities in the 1950s and 1960s, particularly in the U.S., ceased to regard infectious diseases as one of the major causes of death and morbidity, and argued, furthermore, that fundamental research on microorganisms could be halted altogether (Burnet 1953 in Fantini 1993). This perspective was also reflected at the political and economic levels. After the "war" on cancer and cardio-vascular diseases was declared in the early 1970s, for instance, the budget of the National Institute of Health (NIH) doubled in 5 years, while the funding for the National Institute of Allergy and Infectious Diseases (NIAID) grew by only 20 % (Krause 1998, 3). The belief in the power of medical technology to conquer infectious diseases with newly developed drugs resulted in the idea that given sufficient time most of these diseases would naturally decline as a result of the evolutionary dynamics that govern host and pathogens' relation and lead to lower levels of virulence over time (Méthot 2012a; Snowden 2008).

The return of infectious diseases from the early 1980s onwards turned this perspective on its head, however, as the responses of modern medicine seemed no longer adequate in the face of the steep rise of nosocomial infections and the evolution of drug resistance worldwide. Particularly, the acute sense of control over

---

[1] Here we use the term 'parasite' in its broad (ecological) sense, which encompasses both micro-parasites (viruses, bacteria, protozoa) and macro-parasites (e.g. worms).

[2] On the concept of pathogen, see Méthot, P.O. and Alizon, S. (forthcoming).

[3] For classic statements of a natural decline of infectious diseases, see for instance Cockburn (1963); Burnet (1946). For a critical review, see Ewald (1994), and for a historical account, see Méthot (2012a).

infectious disease felt by many was thrown into disarray with the onset of the HIV pandemic and other emerging infections such as Ebola fever, SARS, and more recently with the return of H1N1 influenza. Partly because "many people find it difficult to accommodate the reality that Nature is far from benign" (Lederberg 1993, 3), the rationale of the "conventional wisdom" (as named by May and Anderson 1983) – namely that hosts and pathogens should coevolve towards a state of harmlessness – was promoted far into the second half of the twentieth century (see Ewald 1994 for a review). An additional reason for the success of this avirulence hypothesis, besides its intuitive soundness, was the fact that no serious alternatives to it were introduced before the late 1970s (Alizon et al. 2009), even though some like zoologist Gordon Ball (1943) did raise important objections to the conventional wisdom. The thesis of a natural decline in the virulence of infectious disease postulated by earlier evolution-led models has been challenged on both theoretical and experimental grounds in the last 30 years. Empirical evidence and advances in modelling in evolutionary ecology (e.g. the trade-off model) have shown, for instance, that the evolution of hosts and parasites into a commensal state is not the vanishing, obligate point it was once held to be, but is rather only one of the possible evolutionary outcomes (Anderson and May 1982; Levin and Pimentel 1981; Ewald 1983; reviewed in Alizon et al. 2009). As biologist Carl Bergstrom recently stressed: "we cannot count on evolution to do our work for us" (Bergstrom 2008, 261). Selective pressures, on the contrary, can drive the emergence of new diseases (Antia et al. 2003). And as some have argued, humans may well be the "world's greatest evolutionary force" (Palumbi 2001) behind the increased virulence of pathogens.

Through new social and cultural practices we open-up new routes for "viral traffic" (e.g. blood transfusions, organ transplants), foster behavioural changes facilitating pathogens' transmission (e.g. air travel, migrations, sexual practices, use of drugs, etc.), and introduce "new" pathogens from different parts of the world into immunologically naive populations (Morse 1991, 1993, 1995). This of course adds up to the continuing emergence of human pathogens through zoonotic reservoirs (Wolfe et al. 2007). Infectious diseases continue to be a serious threat to human health, and some diseases once believed to be eradicated might return. Between 1940 and 2004, 335 diseases have emerged in human populations, the majority of them appearing during the 1980s after rapid increase in drug resistance was detected (Jones et al. 2008). Despite the recent steep rise in chronic and degenerative illnesses, emerging infections are still a global challenge for twenty-first century biomedicine and they continue to claim 15 million lives annually (Morens et al. 2004; Fauci 2000). Following the resurgence of infectious diseases as a leading cause of death and morbidity, and the detection of previously unknown disease-causing entities, the idea that newly emerged pathogens have thrown the natural world "out of balance" (Garrett 1994) has garnered a significant amount of scientific attention and has led to the adoption of new international health regulations in order to monitor, limit, and control the spread of communicable diseases (Castillo-Salgado 2010). Here, we explore how, and in what contexts (molecular, ecological, and evolutionary), knowledge claims about disease emergence and changes in virulence are made and justified in the case one specific example: the 1918–19 influenza pandemic.

Emerging diseases are usually defined as diseases whose incidence has significantly increased within a population over a definite period of time (Morse 1995).[4] As Weir and Mykhalovski (2010) recently observed, two of the most influential books on emerging diseases in the early 1990s (Lederberg et al. 1992; Morse 1993) have stressed the need to investigate factors driving disease emergence from both an ecological *and* a molecular-genetic point of view. Both books argued that the biology of the host and the pathogen, in addition to their complex interactions in changing ecological and evolutionary contexts, must be carefully considered in order to devise appropriate public health measures. In practice, though, it remains a challenging task to integrate those perspectives. Indeed, our starting point is the current gap – and lack of integration – in the literature between studies of virulence as applied to emerging disease in the biomedical sciences broadly understood and in molecular pathology and evolutionary ecology in particular. Integration is a multi-faceted concept that is often promoted as a promising goal of scientific practice. As discussed by philosophers of science, integration in science is a complex process that encompasses several activities such as methodological integration, data integration, and explanatory integration (O'Malley and Soyer 2012; see also Mitchell 2008), among others.[5] More rarely is the possibility that integration will fail discussed, however (see O'Malley (2013) for an example of such). As this chapter exemplifies, ecological and molecular methodologies have yet to come together to provide a broader picture of changes in virulence in emerging diseases. Here, we focus particularly on experimental and modelling practices in molecular biomedicine and evolutionary ecology and on their respective explanatory limitations. Very often, explanations of the virulence of a pandemics are constructed as an alternative between knowing the biological nature of the pathogen or that of the environmental conditions that facilitate its transmission. While both consider the nature of the host as part of the disease process, most of the time one branch of the alternative alone is considered as the right (or at least sufficient) explanation while attention to other explanatory schemes is scant. Using the 1918–19 influenza pandemic as a case study of a particularly virulent emerging disease, we illustrate the enduring persistence of two distinct scientific styles of practice in the recent history of virulence studies.

Beginning with a discussion of the evolution of virulence as seen through the lens of ecological and molecular perspectives in biology, we show how each of them conceptualizes both the nature of virulence and emergence in quite different ways. Next, we describe the biology of influenza viruses with a focus on the 1918–19 pandemics and we move on to the ecological-evolutionary explanations of its exceptional virulence, paying attention to the trade-off model, before turning to molecular

---

[4] On the history, epistemology, and social aspects of the concept of emerging disease see Grmek (1993); Farmer (1996), King (2004); and Weir and Mykhalovski (2010).

[5] See the recent special issue on integration in *Studies in History and Philosophy of the Biological and Biomedical Sciences* (2013).

pathology. We argue that an evolutionary approach is necessary to understand the dynamics of disease transmission and evolution but that a broader understanding of virulence will ultimately benefit from articulating the ecological dynamics with cellular mechanisms of virulence. In sum, both ecological and functional perspectives on host-pathogens' interactions are required to answer the opening question of this essay but also to devise appropriate health-care measures in order to prevent (and predict?) future influenza pandemics and other emerging threats. The difficult co-existence of distinct explanatory frameworks reflects the fact that scientists can work on a same problem using distinct methodologies (Godfrey-Smith 2006), but it also highlights the enduring tension between two scientific styles of practice in biomedicine.

Functional and Ecological Perspectives on Emerging Diseases and Virulence

Evolutionary biologist Ernst Mayr has long suggested that functional (proximate) and evolutionary (ultimate) perspectives in biology lack unification (Mayr 1961; see Morange 2005). More recently, evolutionary ecologists have argued in the direction of a better integration of those perspectives (Frank and Schmid-Hempel 2008). While Mayr's point that proximate and ultimate explanations are not alternatives is sound, developmental biology advocates, among others, have persuasively argued that evolutionary questions are relevant to understanding developmental processes, and vice-versa (see Laland et al. 2011 for a review). Today, another, and perhaps equally significant divide, seems to be that between ecological and functional (or proximate) approaches to biological systems and their evolution. As we show, what we call exogenous and endogenous approaches to virulence both make knowledge claims based (sometimes loosely) on evolutionary theory, although each of them invokes one particular aspect of the theory.[6] Whereas the ecological (or exogenous) style focuses on processes (e.g. selective pressures, population density, within and between host competition, and so on) acting on the hosts and the pathogen, the molecular (or endogenous) style traces the evolutionary pathway, or patterns, of the influenza virus from animal(s) to man, and, by constructing molecular phylogenies, identifies particular genes for pathogenesis and mutation sites within lineages.[7] In other words, the former analyses one of the main mechanisms of evolution (i.e. natural selection) and the latter describe the path of evolution (i.e. they construct phylogenies) (Ruse 1992). The construction of molecular phylogenetic trees by

---

[6] The use of the concept of "exogenous" and "endogenous" styles is inspired by the work of historian of science Ton van Helvoort (1994). In turn, this approach is indebted to Polish immunologist and epistemologist Ludwik Fleck (1979).

[7] Patterns derive from processes. The former can be described as the "study of order in nature" while the second refers to "mechanisms generating and maintaining this order" (Chapleau, Johansen, and Williamson 1988, 136).

molecular pathologists reflects the recent "data-driven" trend itself supported by genomics, molecular biology, and the development of high throughput technologies. The use of "evolution" by molecular pathologists is, however, secondary to finding molecular mechanisms for pathogenesis and thus explaining changes in virulence mechanistically.

Each perspective also provides a different way of thinking about disease emergence. Briefly, the endogenous view describes how bacteria and viruses can be transformed into pathogenic, emerging diseases by gaining intracellular and genetic material such as, for instance, a polysaccharide capsule, a large plasmid, a set of virulence genes, or pathogenicity islands (Friesen et al. 2006). These and similar findings have led some to claim that pathogens can evolve in "quantum leap" (Groisman and Ochman 1996). Point mutations allowing the virus to bind to a host receptor also belong to this category. While the capacity to cause disease due to new sets of genes is a crucial aspect of how organisms become pathogenic, this capacity can also occasionally result from genomic deletion and gene loss (Maurelli 2006). In sum, acquisition of novel "virulence factors" (or deletion of other genetic elements) can rapidly lead to the emergence of new diseases or enhanced virulence in some pathogens. For molecular pathologists the concept of virulence is similar to the traditional definition of plant pathologists, i.e. the infectivity: a strain is virulent if it is able to infect a host.[8] This definition could be traced back to the work of Pasteur, for whom "virulent cultures killed, attenuated ones did not" (Mendelsohn 2002, 3–26, p. 5). A more classical definition is the ability to generate symptoms. In both cases, virulence is an all or nothing trait; it is qualitative and not quantitative. Note that these definitions have the advantage that they can be translated at different levels, for instance at the cellular level, where virulence can be the ability to infect cells.

The ecological or exogenous style adopts another approach to disease emergence, virulence, and evolution. Often described as a two-step process, disease emergence requires the introduction of a pathogen within a population followed by its successful dissemination (Morse 1991, 392–3). The "rules of viral traffic" (Morse 1991) dictate that both steps usually result from one or several changes in the environment, not from a modification in the biological characteristics of the pathogen. For instance, in 1976 a change in the air conditioning system in a hotel in Philadelphia facilitated the spread of Legionellosis, a bacterium usually commensal to humans, which caused an outbreak of fever and pneumonia now known as Legionnaire's disease. However, there are cases supporting a biological explanation of emergence, for instance when a maladapted strain mutates into a well-adapted strain before going extinct (Antia et al. 2003). The trade-off model developed by Robert May and Roy Anderson, and independently by Paul Ewald, in the early

---

[8] Note that, for historical reasons, in the phytopathology literature the virulence used to refer to the ability of the pathogen to infect a plant (i.e. a qualitative trait). Since 2001 the American Phytopathological Society has decided to use the term virulence to refer to the damage done to the host and the term pathogenicity for the ability to infect the plant but few researchers have adopted it. In a way, the debate between two fields (evolutionary ecology and molecular biology) has already happened within one of the fields (see Shapiro-Ilan et al. 2005; Thomas and Elkinton 2004; Shaner et al. 1992).

1980s currently underpins the bulk of the theoretical research on host-pathogen's interactions in evolutionary ecology.[9] Put simply, the model postulates the existence of ecological trade-offs between a number of epidemiological variables. As a consequence, the evolution of virulence becomes linked to several factors: host resistance and recovery rate, pathogen transmission rate, the timing of infection life-history events and population density, among others. The trade-off model permits the investigation of the role of environmental changes broadly conceived (including within and between hosts selection) and selective pressures acting on pathogen transmission, and thus on the level of virulence (Alizon 2014).

While molecular geneticists quickly adopted the concepts of virulence genes and pathogenicity islands, evolutionary ecologists working with the trade-off model continued to regard them with suspicion (see Poulin and Combes 1999).[10] We think this suspicion is probably due to the way virulence is defined. For evolutionary biologists, virulence typically is a quantitative trait that can be measured. Therefore, genes that are sufficient to render a pathogen virulent and essentially act as a qualitative trait are difficult to fit into the picture. Furthermore, there is no such thing as pathogen virulence alone in ecology. Virulence, typically, is a "shared trait" that results from the interaction between a host genotype, a parasite genotype and their environment. In other words, some parasite genotypes might only cause virulence when they infect some host genotypes or some parasites may only be virulent to hosts in certain contexts (e.g. starvation). For evolutionary biologists and ecologists, virulence is the harm a pathogen does to its host, i.e. the reduction in host fitness due to the infection (Read 1994). Fitness is notoriously difficult to evaluate but arguably the two most common measures are lifespan and fecundity.[11] One problem is that a pathogen strain described as being very virulent in vitro could turn out to be mild in vivo (and vice-versa). Furthermore, recent work shows that levels of virulence can actually be the result of the immune system's over-response itself (see Graham et al. 2005 for a review). In the end, evolutionary ecologists focus on a combination of within-host processes when they refer to virulence. Importantly, this does not mean that they disregard the molecular processes that lead to virulence. For instance, studies have shown that immune-pathology contributions to virulence lead to a different evolutionary outcome than "virulence factors" produced by pathogens (Alizon and van Baalen 2005).

Recent explanations advanced to account for the rapid changes in virulence during the 1918–19 influenza pandemic reflect the polarity between ecological and molecular explicative strategies. Applying the trade-off model to the 1918–19 pandemic, Paul Ewald has argued that the proximity of soldiers in the trenches, the hospitals, the transport, and the military camps during World War I greatly facilitated transmission of the virus from host to host. High viral replication rate by

---

[9] On the origins of the trade-off model in May's work, see Méthot (2012a).

[10] Before the formulation of the trade-off model, Macfalane Burnet declared: "there are no virulence genes as such" (1960, 1).

[11] Survival and reproduction often interact in a non-linear way. For a discussion on the epistemological aspects of the concept of fitness, see Bouchard (2004).

natural selection was therefore favoured, which resulted in exceptionally high virulence and the high level of mortality of the pandemic (Ewald 1991, 1994, 1996). But
since the late 1990s, molecular pathology has provided an alternative viewpoint on
the evolution of virulence in the pandemic. The identification of the viral RNA from
frozen bodies and wax blocks in the U.S. and its further sequencing has led to a
renewed emphasis on genetic and molecular determinants of the virus as being the
most important cause of this dramatic event (see Holmes 2004). According to
molecular pathologist Jeffrey Taubenberger, one of the leading scientists involved
in reviving the 1918 influenza strain, "it is possible that a mutation or reassortment occurred in the late summer of 1918, resulting in significantly enhanced
virulence" (2005, 90). Taubenberger believes that this "unique feature" of the
1918 virus – its extreme virulence – "could be revealed in its [genetic] sequence"
(2005, 90). Both approaches – the exogenous and the endogenous – evolved along
parallel lines during most of the twentieth century, and though the concept of
emerging infectious diseases brought them closer to one another in the 1990s, we
show how they remain in tension (Méthot 2012b). Before describing in more
details the potentials and limits of these two perspectives we first describe significant aspects of the biology of influenza viruses.

The Biology of Influenza Viruses and the 1918–19 Pandemic

The natural history and ecology of influenza A virus has been extensively studied
(Webster 1999, 1993; Webster et al. 1992; Webster and Rott 1987). The virus'
natural reservoir is the wild waterfowl, as supported by the fact that species of wild
duck are not affected by the virus and remain "healthy". The virus replicates inside
the host, mostly in the intestinal tract, and is then washed into the ponds where
ducks live and breed (Webster 1993). The relative harmlessness of this relationship
is similar to the way myxoma virus is adapted to its natural host, the South
American rabbit (see Fenner and Fantini 1999). The family tree of influenza viruses
contains two genera: one that includes influenza A and B viruses and the other
influenza C viruses. The two genera are distinct in terms of host range and virulence factors. Type A is the most common of all, and can infect a wide range of
hosts, including, pigs, horses, seals, whales and birds. This type of virus is also the
most redoubtable as it has the potential to cause pandemics. Type B is believed to
infect only humans (especially young children) and Type C (another genus) can
infect both humans and swine. In this sense influenza can hardly be regarded as a
"single disease" (Johnson 2006, 10).

Influenza viruses are enveloped negative strand RNA viruses and belong to the
genus *Orthomyxoviridae* (Taubenberger 2005). The virus of the Spanish flu pandemic belongs to the type A influenza, known as H1N1. Influenza A and B viruses
contain eight discrete gene segments, coding for at least one protein. The surface of
influenza A viruses is covered by three types of proteins hemmagglutinin (HA), neuraminidases (NA) and matrix 2 (M2). The structural configuration of HA proteins is
that of a triangular spike. These spikes allow the virus to bind to red blood cells by

causing the latter to agglutinate (i.e. hemmagglutinin). They facilitate entrance into the host and they trigger the infective processes. Once the infection is over, antibodies responding to hemmagglutinin spikes are formed, allowing the immune system to recognize the signature of the viral strain in case of another infection episode. In contrast, neuraminidases (NA) also form spikes on the surface coat of the virus but their function is to cleave glycoproteins into two and to facilitate the propagation of the virus from cell to cell. NA proteins open-up cells for infection, so to speak. Antiviral drugs target NA in order to block their exit, and antibodies to NA are also produced after the infection. Influenza A viruses are further subdivided into serological types, which is the genetic characterization of the surface glycoproteins HA and NA. 16 HA and 9 NA proteins have been described to date. These surface glycoproteins define the virus' identity in terms of what the immune system detects and attacks. The different major families of flu are combinations of the two, hence the designation "H5N1" for the recent threat. The 1918 virus was H1N1.

The genes coding for these glycoproteins can reassort (i.e. reshuffle) due to two processes known as antigenic drift and antigenic shift. The former consists in the accumulation of point mutations in the genome of the virus, modifying both the shape and the electric charge of viral surface antigens and preventing their recognition by the antibodies of the host that were developed in reaction to previous exposures to the virus. The need to update the influenza vaccines every year illustrates the evolutionary success of antigenic drift. In contrast, antigenic shifts refer to the introduction of whole or part of influenza genes into viruses that circulate among human populations. This form of genetic reassortment or reshuffling occurs especially in swine that act as "mixing vessels" for the viral strains and are considered the intermediate host between birds and humans (Webster and Kawaoka 1994). The introduction of a new hemmagglutinin gene (HA) is often hailed as the responsible factor for increased virulence (Bush 2007). The fast reassortment of nucleotides and the high rate of mutation in influenza viruses result in influenza posing a continual threat for human and animal health. As a result, influenza is regarded as being a continually "re-emerging" disease (Webby and Webster 2003; Webster and Kawaoka 1994), and international efforts are made to understand why the 1918–19 pandemic was so exceptionally virulent. The motivation behind these global efforts in gaining a better understanding of this pandemic is to draw lessons from the past in order to be better prepared for the rise of future influenza and other viral pandemics.

Recorded history suggests that the first influenza pandemic occurred in 1580. Beginning in Asia, it rapidly spread to Africa, America and to Europe. Between the eighteenth and the nineteenth century, medical historians identified (at least) 8 pandemics out of 25 epidemics of influenza A virus (Beveridge 1992). The most devastating pandemic, however, occurred in 1918–19 (Fig. 1).[12] The emergence of

---

[12] On the history of the 1918 influenza pandemic, see Barry (2004a), Johnson and Mueller (2002), van Helvoort (1993), Crosby (1989), and Burnet and Clark (1942). For a short but informative "chronicle" of influenza pandemics, see Beveridge (1992), and for a detailed scientific account of the biology of influenza see Stuart-Harris (1953).

For recent histories on flu, see Bresalier, M. 2013 and Taubenberger and Morens (2010).

For a recent account of influenza pandemics, see Honigsbaum, M. A. 2013.

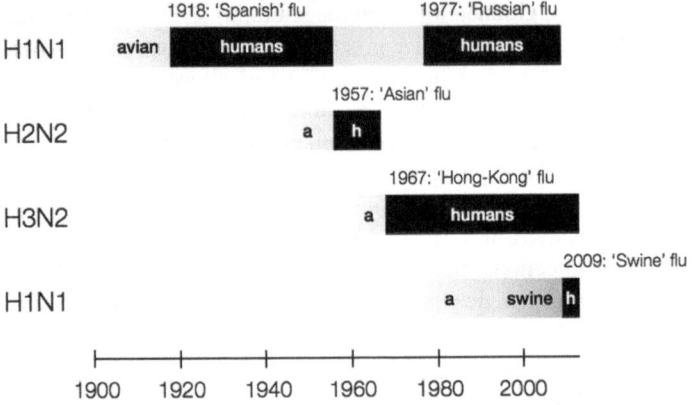

Fig. 1 Human influenza A pandemics and viruses in the twentieth century

the (misnamed) "Spanish" influenza pandemic of 1918–19 is the first of the three major influenza pandemics that occurred during the past century – and is regarded as one of the most devastating episodes in medical history (McNeill 1976).[13] Once described as "the biggest unsolved problem of theoretical epidemiology and public health practice" (Burnet and Clark 1942), its consequences rendered many wary about the emergence of respiratory disease pandemics in a near future (Webby and Webster 2003). In addition to the 1918–19 pandemic, two other major influenza pandemics occurred in 1957–58 ("Asian" influenza, H2N2) and in 1968 ("Hong Kong" influenza, H3N2). The death of David Lewis, a soldier at the military camp of Fort Dix in the U.S., and the infection of a few hundreds of others in 1976 led public health authorities to believe they were facing a new influenza epidemic. Amidst some scepticism, vaccines against H1N1 flu were quickly stockpiled as President Ford gave the green light to mass vaccination. However, no epidemic occurred while a number of vaccinated individuals came down with Guillain-Barré syndrome, an autoimmune disease, a few weeks later (see Krause 1998). One year later, in 1977, the H1N1 virus, which had disappeared in 1957, reappeared (the so-called "Russian" influenza) in the Soviet Union and spread to Taiwan, the Philippines, Singapore, and within 10 months had reached South America and New Zealand. The virus was similar to a virus isolated in the U.S. in 1950 at Fort Warren and had perhaps been accidentally released from a laboratory located in the former

---

[13] The reference to Spain is due to the fact the publication of medical reports on influenza was authorized in Spain during the war, in contrast to other countries at war. As a consequence, the disease became associated with Spain that was subsequently blamed for it and considered responsible. One of the first papers to appear in *London Times* (June 1918) was titled "The Spanish Influenza – a sufferer's symptoms" (in Johnson 2006, 37). Like syphilis, the Spanish influenza received other names in other countries, however. For instance, it was called the "Swiss flu" in France.

Soviet Union (Berche 2012, 127). Affecting mostly individuals born after 1957, this virus coexisted with the H3N2 virus until 2009, when a new variant of H1N1 emerged (the "Swine" flu, which is the latest pandemic) that replaced the 1977 variant. In comparison to the Spanish flu pandemic of 1918, the Hong Kong and the Asian pandemics were more "benign", the former causing between 1.5 and 2 million deaths, and the latter 1 million. The recent H5N1 pandemic caused a few deaths only between 1997 and 2004 (Taubenberger 2005, 87). Despite the (crucial) facts that antibiotics were available during the second two pandemics, and that medical care had significantly improved and was more efficient after 1950, this raises the question: why was the 1918–19 pandemic so deadly to humans?

Three Significant Aspects of the 1918–19 Pandemic

A Western Origin

A first important aspect of the 1918–19 influenza pandemic is its likely Western origin. In part because of its extensive pig-duck farming industry, China was previously singled out as the possible origin of most influenza pandemics. However, whereas most pandemics to have befallen man have come from China (Morse 1993, 17) the "Spanish" flu originated (likely) from France as early as 1916 causing acute respiratory symptoms closely resembling the phenotype of the disease during the 1918–19 flu pandemic (Oxford et al. 2002, 2005). Some have recently argued that there was an early wave of influenza in New York between February and April 1918 (Olson et al. 2005). The precise geographical origins of the 1918 pandemic are still a matter of debate, however.[14] The world's deadliest flu pandemic kicked off in October 1918 and in just a few months, the virus killed between 30 and 40 million people (Philips and Killingray 2003; Johnson and Mueller 2002; Crosby 1989; some estimate deaths to number about 50 million, see McNeil 1976). According to the "three waves theory", influenza swept through all five continents in three recurrences. The first wave (or the "spring wave") of the flu started in March in the U.S. (Mid West) before moving to Europe, then to Asia and North Africa before reaching Australia in July 1918. While morbidity was high, mortality was not higher than the habitual norm (Reid et al. 2001, 81). The second wave (or "fall wave"), however, was highly devastating and rapidly went extinct after causing millions of deaths worldwide, with peaks in October and November. It started in late August 1918 and within 1 week reports of the virus came from distant cities, including Boston (U.S.), Freetown (Africa), and Brest (France). On many accounts, this second wave lasted until November. The speed at which the virus circulated makes it difficult to pinpoint one specific location as being "the" source of the pandemic but a Western

---

[14] For instance, Langford (2005) argues that the flu pandemic came from China.

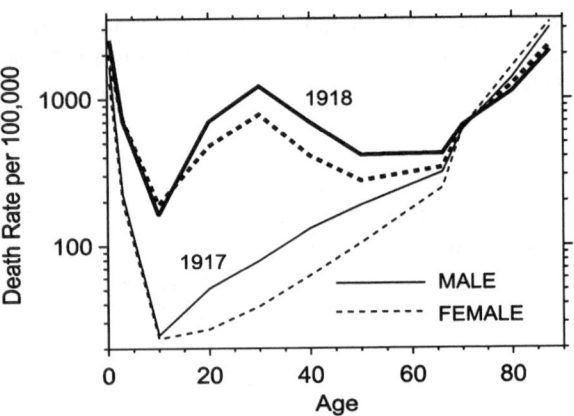

Fig. 2 Age distribution of death rates from influenza and pneumonia in the United States death registration area, 1917 and 1918. Death rate is deaths per 100,000 person-years lived. Data from US Department of Health, Education, and Welfare 1956. In Noymer (2010, 141)

origin appears to be the most plausible hypothesis according to the available evidence. Reports indicated a further third wave that hit in the first months of 1919 but was much less severe (Burnet and Clark 1942; Barry 2004a). However, the three waves pattern of the pandemic is not uniformly applicable to all countries; for instance Australia experienced a single occurrence of the flu pandemic (Johnson and Mueller 2002; Morens and Taubenberger 2009).

Signs, Symptoms, and the Age Group of the Victims

In 1918, during an attack of influenza most victims died of secondary infections as death often resulted from bacteria invading the lungs of immunocompromised individuals (Burnet and Clark 1942). Symptoms lasted generally between 2 and 4 days and could, more rarely, be extended up to 2 weeks. The respiratory disease was characterized by fever, body pain, and severe headaches. Without the possibility of treating patients with antibiotics, bacteria turned "those vital organs [lungs] into sacks of fluids [...] effectively drowning the patient" (Philips and Killingray 2003, 5). People therefore died within just a few days of hemorrhagic pulmonary oedema and other lung afflictions (Bush 2007; see also Taubenberger et al. 2000). Related to this, the second striking aspect of the 1918 influenza pandemic is the young age of the victims, which was qualitatively distinctive: most of them were men, supposedly healthy, of between 20–40 years old (some say 25–35), irrespective of whether the country was involved in the war or not. Instead of forming a U shaped mortality curve, the shape of the 1918 pandemic was W shaped. An additional peak (the central peak in the W) represents the male victims of the flu. Figure 2 (above) shows the U shaped curve of 1917 and the W shaped one of 1918. The distribution of deaths on this curve reflects the virulence of the pandemic and underlines the pattern of mortality of a group usually not affected by seasonal flu.

Exceptional Virulence

Finally, the third and most significant feature of the 1918 pandemic was its lethality: the disease was of exceptional virulence and estimates suggest that the pandemic claimed more victims than the First World War (McNeil 1976). This central aspect was almost universally recognized as being somewhat unusual and very specific to it although the estimates of fatalities during the twentieth century vary between 20 and 50 million deaths (Johnson and Mueller 2002).

Influenza type A viruses are not at all uncommon, and strains had circulated in human populations for a few centuries before since a few centuries when the 1918 pandemic broke out. In the United States alone only, annual death tolls related to seasonal influenza are estimated to be about 30,000 individuals (Thompson et al. 2003). Seasonal outbreaks of influenza normally last a few weeks and then disappear abruptly; they result from influenza viruses present in human populations that are able to infect individuals due to antigenic drift. On occasions, however, the virus can infect up to 40 % of the world population. During these pandemic years, in contrast, the number of deaths rises way above the average, claiming millions of victims all around the globe. In the course of seasonal epidemics strains of influenza type A and B can sometimes coexist, if at different frequencies among populations. So why was the Spanish influenza so devastating? Recent work in molecular biology argues that the waves pattern, the group mortality, and the clinical course of the disease "may find their explanation in genetic features of the 1918 virus" (Reid et al. 2001, 86). Others, however, defend the view that the changes in virulence result from significant changes in the wider ecological context in which the outbreak occurred (Ewald 1994). In the next sections, we review both ecological-environmental and molecular-led approaches to this problem, we indicate some of the limitations of each and we suggest that a better integration of those perspectives would lead to positive outcomes regarding prediction, prevention, and preparedness in the face of other similar influenza and other bacterial or viral pandemics.

Evolutionary Epidemiology and Environmental Explanations of Disease Emergence

From an ecological point of view, for a disease to emerge in a population the basic reproductive rate of the pathogen (R_0) must be higher than 1, where R_0 is the average number of secondary infections that follow from one infected individual in a wholly susceptible population (Anderson and May 1991). In other words, a pathogen must cause at least one subsequent infection to persist in the host population. The classical formula used to capture the trade-off model is as follows:

$$R_0 = \frac{\beta S}{\mu + \alpha + \gamma}$$

where R_0 serves as a measure of Darwinian fitness of the pathogen at the epidemiological level. In the denominator are α, the host mortality due to the infection (i.e. the virulence), μ, the rate of microparasite independent-mortality and γ is the rate of recovery of the host. The inverse of $\mu + \alpha + \gamma$ is the average duration of the infection. In the numerator, we have β, the transmission rate, and S, the host population size. Importantly, one should not confuse β, which is a rate (number of infections per unit of time and per susceptible host in the population) and R_0, which is roughly the number of new hosts infected over the whole duration of the infection. Overall, this expression indicates that parasite fitness is the product between the number of secondary infections generated per unit of time and the duration of the infection. Any animal that produces less than one offspring over its lifetime infections generating less than 1 secondary infection will eventually become extinct and die out.

This $R_0 > 1$ threshold is of course a simplification. For instance, in the early stages of an outbreak, emerging pathogens infect very few hosts which means that they are particularly prone to extinction going extinct due to stochastic effects. In fact, it can be shown that in an ideal situation where all the hosts would be identical, the probability of emergence of a pathogen with an R_0 strictly greater than unity is only of $1-1/R_0$, due to these stochastic effects (Diekmann and Heesterbeek 2000). Conversely, pathogens with $R_0 < 1$ can nevertheless be dangerous because they can persist in the population for a while stochastically, which leaves time for a variant with an $R_0 > 1$ to evolve (Antia et al. 2003). In other words, the transmission between hosts (in these cases humans) must be effective for disease emergence to occur. Historically, the first example of a trade-off came from the analysis of myxoma virus infecting rabbits (Anderson and May 1982; Fenner and Fantini 1999). However, since then, clearer examples have been worked out. Fraser et al. (2007), for instance, combined data on virulence from an Amsterdam cohort and data on transmission rate in discordant HIV-infected couples from a Rakai cohort to show that individuals with a higher set-point viral load (i.e. the viral load during the asymptomatic stage, which has the property to often remain constant over several years) have a shorter lifespan and a higher transmission rate. When they combined host lifespan and virulence together to obtain a measure of parasite fitness (i.e. R_0), they found that viruses with an intermediate virus load achieved the highest fitness. They also show the observed abundance of virus loads in a human population follows the distribution virus fitnesses.

The classic trade-off model focuses primarily on pathogen populations and their evolution. It often ignores host evolution because generation times for hosts (here, humans) tend to be much longer and so evolution in the host population is likely to be slow. From a trade-off model perspective, pathogens' rate of replication within a host, which usually increases the probability of its being transmitted to a new host, is balanced with its negative effect on the duration of the infectious period (May and Anderson 1983). If the pathogen is not virulent, it is unlikely to reach a high transmission rate. Conversely, a pathogen that replicates intensively in the host will have a higher transmission rate but over a shorter time because rapid host exploitation also means shorter host lifespan. Similarly to Achilles who, according to Homer, had to choose between a short and glorious life or a long but dull one, the pathogen has to evolve a strategy between causing long and mild infections or short and virulent infections (Alizon et al. 2009). If such a trade-off is at work, external factors can affect virulence evolution in a predictable way. For instance, the lower the baseline mortality of the host (independently of the disease), then the higher the pathogen virulence should be. This is so because the infectious period is reduced and the pathogen has to use up the host resources in a shorter time. There is also a growing interest in the host reaction to an infection, which can broadly be split into resistance (i.e. fighting the disease, which decreases both virulence and transmission) or tolerance phenomena (decreasing only the virulence, not the transmission rate) that affect parasite evolution (Raberg et al. 2009; Boots et al. 2009).

The density of susceptible hosts can also affect short-term evolutionary dynamics of virulence, as clearly shown by a recent evolutionary epidemiology framework that combines epidemiology and population genetics (Day and Proulx 2004). Indeed, early on during the course of an epidemic, most of the hosts are susceptible to infection so natural selection acts to favour strains with a high transmission rate (which happen to be more virulent according to the trade-off hypothesis). Once the disease has reached an endemic stage, the pool of susceptible hosts is smaller (hosts are either already infected, dead, or immunised) and natural selection then acts to favour strains that cause longer infections. In conclusion, virulence can thus be expected to vary over the course of an epidemic for rapidly evolving pathogens.

Another dimension of the model is that it does not concern itself with morbidity (at least not explicitly). Thus, symptoms like pain or injuries are not taken into account by the trade-off model and are implicitly integrated with other variables like host recovery and parasite transmission (Levin 1996). This assumption impacts on the ways in which virulence will be measured and operationalized. Whereas for doctors morbidity (illness) is a key feature of virulence, for evolutionary biologists or population biologists host's pathological factors do not need to be taken into account when measuring virulence; what matters are effects that modify the pathogen's fitness (i.e. that appear in the expression of R_0). In sum, the model rests on the idea that the pathogen transmission rate cannot increase beyond a certain point without at the same time inflicting damage to the host which would, in turn, be harmful to the pathogen by decreasing the duration of the infection. What matters for an evolutionary biologist is the fitness of an individual where the fitness of a parasite strain typically is given by the R_0, i.e. the number of secondary infections. In other words, for a given parasite species, natural selection favours strains with the highest R_0. This can explain why the highest possible level of virulence is not always the evolutionary stable ("optimal") strategy to increase parasite's fitness: increased transmission will indeed increase one component of parasite fitness R_0 (the transmission rate) but it will also decrease another component of R_0 that is the duration of the infection (through increased virulence) as the host is likely to die more rapidly from the infection. The balance between the two selective pressures (transmission favouring higher virulence and duration of infection favouring lower virulence) determines the evolutionary stable level of virulence.

Applying the Trade-Off Model to the 1918–19 Influenza Pandemic

Ewald's early work on pathogen's virulence and transmission developed a verbal theory for the trade-off model by comparing diseases with different transmission routes (Ewald 1983). His work was based on the concept of "cultural vectors" and on the assumption that parasites that do not rely on host mobility for transmission should evolve towards higher levels of virulence. In Ewald's terminology, a cultural vector is "a set of characteristics that allow transmission from immobilized hosts to susceptible when at least one of the characteristics is some aspects of human cultures" (1994; 68; see also Ewald 1988). In the case of waterborne transmission, cultural vectors include contaminated bed sheets in hospitals, sewage systems carrying the pathogens, medical staff disposing of the contaminated water to water supplies, and so on. Waterborne diseases can become more virulent because they do not rely on host mobility for transmission (see Ewald 1994, 69), that is, the host can be isolated and still be highly contagious; a "healthy" host is not needed for transmission (in contrast with what was postulated by the conventional wisdom). Note that implicit in his reasoning is the idea that more virulent pathogens have a higher transmission because they produce more spores.

Applying the trade-off model to the case of the 1918 pandemic, Ewald argued that host proximity and population density were key elements in enhancing virulence. More precisely, he argued that the exceptionally high virulence resulted from rapid passages of the virus in soldiers, recruits and wounded people in hospitals during the war. Though a similar explanation had already been heralded in the 1930s–1940s, it had to be supplemented with an essential "evolutionary mechanism": the classical explanation is based on the analogy with rapid passages of a viral strain through a series of animals (i.e. guinea pigs) in a laboratory that can enhance virulence (as Pasteur et al. (1994 [1881]) had experimentally demonstrated, see Mendelsohn (2002)). Ewald's argument is that, just like biological vectors, cultural vectors enhance virulence by facilitating transmission. The central point about the serial passages is that it removes the "requirement that hosts be mobile to transmit their infections" (Ewald 1994, 115). Once this obstacle is lifted nothing (a priori) stands in the way of a steep increase in virulence. In a laboratory context, experimenters inoculate different animals with artificially selected viral strains; in the field, this selection process results from another cultural vector, namely the warfare conditions.

In the trenches, during the Great War, conditions were such that transmission was maximized and with it, the observed level of virulence. As postulated by the trade-off model, the density of the population (S) influences the level of observed virulence in a biological system (at least for short-term evolutionary dynamics). In this case, the high density resulted from the proximity of the soldiers in the trenches, in hospitals, on trains bringing soldiers to the front, and in military camps. In turn, this resulted in the unusual situation that immobilized individuals who normally should not be able to infect new people (because they would be isolated in a hospital) were now easily able to transmit the infections. Similarly, removing wounded soldiers from the trenches and bringing them to war hospitals facilitated transmission. The constant arrival of new susceptible individuals into the population through transport networks resulted in maintaining a high density of infected people; and as a consequence, an equally high level of virulence. Related to this is the idea that spatial structure in the host population affects virulence evolution. If hosts tend to have few contacts among them, e.g. because they live in isolation (the technical term to describe such a population is "viscous"), then a parasite has to keep its host alive sufficiently long enough to be transmitted. On the other hand, if the population is "well mixed", host encounter rate is not an issue – as in the 1918–19 example – and parasites can afford to be more virulent (Boots and Sasaki 1999).

Some Problems with Ewald's Account

Despite its theoretical appeal, some detected a number of problems in the explanation advanced by Ewald and with the trade-off model in general. For other evolutionary ecologists, Ewald's cost-benefit argument is too adaptationist – i.e. virulence is depicted as being *always* adaptive for the parasite. As a consequence, "alternatives such as virulence being non-adaptive, or virulence being a consequence of short-sighted, within-host evolution of the parasite are ignored" (Bull and Levin 1994, 1470). Evolutionary theory states that virulence can be directly selected but it can also be coincidental with other infection or biological processes (Levin and Edén 1990), and in some cases it can be potentially maladaptive. This point connects to one of the usual critiques levered against the trade-off hypothesis, namely that it is very verbal and lacks empirical support (Levin and Bull 1994; Lipsitch and Moxon 1997). However, the lack of support largely comes from the difficulty of finding an appropriate biological system; arguably, when people have looked for a trade-off in a host parasite system that satisfies the assumptions of the theory they have found it (Alizon et al. 2009). There is actually a tendency to challenge the trade-off hypothesis using host-parasite systems that do not fit the underlying model (see Alizon and Michalakis (2011) for an illustration).

A second problem stems from the low level of transmissibility in influenza viruses and rate of pathogens' reproduction. When taken into account, this concern weakens the claim that high transmission in the case of the 1918 pandemic has favoured high virulence because transmission was lower than with most infectious diseases. R_0 are typically variable but given Ewald's argument it would be expected to find a high transmissibility rate between the virus and its hosts. In turn, this would support the claim that natural selection acted on transmission in ways that increased the overall level of virulence. Moreover, the trade-off assumes a homogeneous population and was developed for diseases transmitted by contact like influenza. However, a comparison of R_0 between the 1918–19 pandemic with other major disease outbreaks in recent history, or with influenza pandemics in general, does not reveal a significantly higher transmissibility in the case of the Spanish influenza.

Calculations suggest that the basic reproductive rate of viruses during influenza pandemics ranges from 2 and 3 (Mills et al. 2004). In comparison, the reproductive rate during an outbreak of measles in England in 1947–1950 was between 13 and 14 secondary infections; a pertussis outbreak in Maryland (U.S.) in 1913 yielded a reproductive rate between 7 and 8; and a mumps outbreak in the Netherlands during the 1970s produced between 11 and 14 secondary infections in a wholly susceptible populations (Anderson and May 1991, 70). In the case of the 1918 pandemic, more recent calculations suggest that R_0 was perhaps equal to 2 (Morse 2007, 7314). Finally, a recent article on the transmission of influenza in households during the pandemic (Fraser et al. 2011) used historical data and mathematical models to study the rate of transmission. The authors found a relatively low level of transmission between individuals and suggest that prior immunity to the virus should be considered. Though transmissibility may, theoretically, have been fostered so that the virus reached unprecedented virulence, the trade-off model alone does not fully explain why it was so deadly.

A third issue is the lack of empirical details in Ewald's explanation of the steep increase in virulence circa 1918–19. To make his argument more compelling, Ewald needs additional data that accurately and empirically describe the environmental conditions in the trenches. For instance, how close were the troops? How many soldiers were there? And more importantly, what was the rate of transmission between hosts? If a similar study to Fraser et al. (2011) could be conducted on viral transmission in the trenches it would perhaps yield interesting insights into the changes of virulence. To date, no epidemiological data exists that could serve as a basis to model the dynamic patterns, however. Though Ewald's account seems to suffer from a number of theoretical and empirical problems, it nevertheless supports the argument that properties other than those of the virus need to be taken into account and that without them we would are not able to fully understand the changes in virulence that occurred. As he remarked, progress towards the evolution of virulence "has largely been limited to improve understanding of the genetic mechanisms of antigenic changes and the influences of these changes and host immunity on the occurrence of epidemics" (Ewald 1991, 15). The recent work of microbiologist John Oxford on what we could call the "War Hypothesis" reinforces Ewald's conclusion by feeding in some of the missing empirical and historical data.

The "War Hypothesis"

While many would agree that the Great War is a variable that must be included, in one way or another, in the broader explanation of the steep evolution of virulence of the 1918–19 pandemic, Ewald is convinced that the influenza pandemic was "caused evolutionarily by the war rather than being just coincidental with the war" (Ewald 1994, 115). The "War Hypothesis", as we may call it, received new support from Oxford (2001; Reid et al. 2003; Oxford et al. 2005) who does not claim that the Great War caused the disease, evolutionarily or otherwise, but instead that the war created the right environment for the virus to become extremely deadly. When the 1918 pandemic broke out air travel was minimal and this suggests, according to Oxford, that "earlier 'seeding' has occurred" (2001, 1857). Taking an environmentally oriented approach to the evolution of virulence, Oxford and his colleagues argued that the 1918–19 pandemic originated in France in 1916 before going global 2 years later.[15] They did not postulate the evolutionary emergence of a mutant strain but rather that the ecological conditions facilitated the spread of a pre-existing influenza strain. Studying several epidemiological and medical reports of sporadic outbreaks of respiratory infections at the British base camp in the town of Etaples in Northern France in 1916, Oxford argued that the disease's clinical picture maps very precisely onto the description of the 1918–19 influenza: not only were the 1916 respiratory diseases extremely deadly, but post-mortem examination revealed in most cases clear evidence of bronchopneumonia and histological analyses of lung tissues indicated "acute purulent bronchitis" (Oxford 2001, 1857).

In an article published a few years later (Oxford et al. 2005), Oxford and his colleagues took their examination of the situation one step further. Rejecting the possibility that "a particular virulence gene of influenza" could help to identify future pandemics, they argued that surveillance and detection of emerging influenza pandemics would be best served by understanding the contexts that give rise to pandemics, rather than by an analysis of genetic factors alone. In particular, concerning the 1918 pandemic, they noted that so far "there is no clear genetic indication of why this virus [the 1918 strain] was so virulent". They also remarked that what is needed is a closer examination of the environmental and social conditions of the time such as population upheavals to explain the exceptional virulence. The authors asked specifically whether "the special circumstances engendered in the war itself have allowed or caused the emergence, evolution and spread of a pandemic virus" (Oxford et al. 2005, 941). For them, the "unprecedented circumstances" of the war in Europe were critical. Back in 1918, the Front was

a landscape that was contaminated with respiratory irritants such as chlorine and phosgene, and characterized by stress and overcrowding, the partial starvation of its civilians, and the opportunity for rapid "passages" of influenza in young soldiers would have provided the opportunity for small mutational charges throughout the viral genome [...] could have been important factors in the evolution of the virus into a particularly virulent form (Oxford et al. 2002, 113).

The military camp of Etaples in France was subject to high traffic in 1916–1917. In addition to soldiers moving up to the Front and back, 230,000 sick and injured individuals were in the hospitals "at any given time", making them overcrowded and allowing the virus plenty of opportunities for "rapid passages". Overall, it is estimated that the region of Etaples hosted two million soldiers who camped there during the war, in addition to the six million others who occupied and fought in the trenches system that connected the English Channel with Switzerland (Oxford et al. 2005, 942). Secondly, as the camp had an "extensive piggery", villagers could buy geese, ducks, and chickens, providing ideal conditions for the influenza virus to

---

[15] This research was later criticized in turn by Barry (2004b) who argued that the influenza pandemic originated in Kansas and was taken to Europe by U.S. soldiers in 1918.

undergo antigenic shift. Thirdly, the extensive use of gases during the war (estimated at one hundred tons), some of which were mutagenic rendered the soldiers immunocompromised and more susceptible to influenza infections. Finally, demobilisation after the war sent the soldiers back home by boat or by train, and contributed to the spread of the disease by person-to-person contact all over the world (Oxford et al. 2005). Taken together, all these factors (overcrowding, being immunocompromised, pig-duck farming, demobilisation) created exceptional conditions for the virus to go pandemic. Ewald had noted that in the absence of a recreation of those circumstances it is unlikely that such a severe pandemic will happen again. What Oxford and his colleagues emphasized in their turn is that the appropriate response to a future pandemic cannot rest of putative virulence genes alone; one has also to consider the context that will allow the virus to spread in a pandemic fashion. At the same time that this ecological perspective was developed, another view on the sources of virulence was well underway in the United States.

Emerging Technologies

In the mid-twentieth century, leading British bacteriologist Wilson Smith, co-discoverer of the viral nature of influenza in humans in 1933,[16] doubted that the exceptional virulence could be linked to a particular genetic or molecular structure of the virus alone: "if we had the chance of getting a 1918–19 strain of the influenza virus now", he said, "it is at least conceivable that, on comparing with the Asian strain, we might find no difference in intrinsic virulence at all, but the conditions in the human population during the two epidemic periods might have affected the degree of heterogeneity displayed by viruses possessed of the same intrinsic virulence" (1960, 77). His comments were intended to provide support to a paper delivered earlier by Edwin Kilbourne, an American virologist who specialised in influenza, who had argued that the greater virulence of the 1918 pandemic was due to a combination of "the emergence of a new antigenic type in a population with little specific immunity" and "the dislocation of and crowding of wartime which favoured not only dissemination and high dosage of virus but spread of bacterial pathogens to an unusual degree" (Kilbourne 1960, 74). Kilbourne argued that the "study of the host and his environment are more crucial to the interpretation of virulence than laboratory study of the virus itself" (1960, 71). Attempts to locate the cause of virulence inside specific genes or to relate them to other mobile, structural elements (i.e. plasmids) were met with scepticism by people like Wilson, Kilbourne,

---

[16] Until the late 1920s and early 1930s, and the work of Olitsky and Gates and Richard Shope, most bacteriologists believed that the causative organism of influenza was the Gram negative Pfeiffer bacillus (*Haemophilius influenza*) isolated by German scientist Richard Pfeiffer in 1892. However, the causal role of this organism in influenza aetiology was also disputed, particularly during the 1918–19 pandemic (Witte 2003). The discover of the viral nature of influenza was made by Smith, Andrewes, and Laidlaw in 1933 (Smith et al. 1933).

and Burnet who were interested in large-scale ecological processes and in the formation of evolutionary equilibriums between hosts and parasites. Ecologically minded biologists were also reacting against the growing place of molecular biology since the 1960s and its reductionist vision of the life process and the life sciences. This is how we can interpret Kilbourne who scornfully remarked that "ironically in this era of molecular biology, the control of no infectious disease has yet depended on understanding its molecular mechanisms" (1977, 1228).

In the early 1950s, scientific expeditions were organized to discover the remains of victims of the Spanish flu in the hope of finding traces of the virus. One of the expeditors was John Hultin (1925-), a pathologist from Sweden who immigrated to Iowa in 1949 to study medicine. As part of a project funded by the University of Iowa, he travelled in 1951 to a small Inuit village whose population was decimated by the 1918 pandemic, which had killed 70 people in a week, a loss amounting to 85 % of the inhabitants. Hoping to find preserved corpses buried in the permafrost hosting traces of the infectious organism, Hultin travelled to the Seward Peninsular of Alaska in a village known as Teller Mission (Taubenberger 2003). He extracted lung tissue from several bodies he exhumed from the village cemetery, but all attempts to culture remaining traces of the virus of influenza from these samples failed to give any result. Forty years later, in the context of the Human Genome Project, the idea of resurrecting the influenza virus surfaced again, this time powered by genomic technology.

Since the late 1990s, a renewed emphasis has been placed on the molecular, internal constituents of virulence. Newly developed technology and the availability of pathogenic viral and bacterial material have facilitated the development of this approach towards explaining infectiousness. This led, in 2005, to the publication of the complete influenza virus' genomic map in both *Nature* and *Science*. Though all samples of the 1918 strain were thought to be long extinct and lost, bits of RNA of the virus were discovered and processed in order to generate the complete map of its genetic structure. After the discovery of frozen individuals killed by the 1918–19 pandemic and preserved in permafrost, scientists worked on the pathogenic mechanisms that possibly enabled the influenza virus to achieve unprecedented levels of virulence. Microbiologist Jeffrey Taubenberger of the National Institute of Allergies and Infectious Diseases in Washington led this work together with Terrence Tumpey from the Center for Disease Control in Atlanta. We now turn to this recent technological success and the difficulties of pinpointing any particular molecular feature of the flu virus of 1918 that could account for its exceptional virulence.

Traces of the Spanish Flu: From (Sero)archeology to PCR Amplification

In 1997 a U.S. lab-group based at the Armed Forces Institute of Pathology in Washington (D.C.), and led by molecular pathologist Jeffrey Taubenberger, published a piece in *Science* titled "Initial genetic characterization of the 1918–19

'Spanish' influenza virus" (Taubenberger et al. 1997). The article provided a first and partial genetic map of the virus from "archival formalin-fixed, paraffin-embedded autopsy tissues of 1918 flu victims" (Taubenberger 2003, 42). The examined samples were kept at, and provided by, the National Tissue Repository of the Armed Forces Institute of Pathology. As several mutations in hemmagglutinin, especially on cleavage-sites, often contribute to the virulence (e.g. on influenza subtypes H5 and H7) by increasing the tissue tropism, it was hoped that the genetic make-up of the virus would provide insights into the virulence of the 1918 Spanish influenza pandemic. The goal of the project was "first, to discover where the 1918 influenza came from, and how it got into people, and second, whether there were any genetic features of the sequence that would give insight into the exceptional virulence of the strain" (Taubenberger 2003, 44).

This first publication of the team describes the technique used to obtain, amplify (PCR), and sequence the genetic material. The main finding of the paper, based on molecular phylogenetic analyses of gene segments, was that the 1918 pandemic was caused by a strain of H1N1 influenza virus, and that it was of avian origin (1997, 1795). In their first article, Taubenberger et al. randomly selected 28 cases of paraffin-embedded tissues collected from army servicemen who died during the pandemic for pathological review, searching for symptoms indicative of death by influenza. Most of the individuals examined died of secondary pulmonary infection, which was a common feature of victims of the pandemic. In effect, bacterial infection very often works together with the influenza virus in delineating the clinical picture of the disease. One case, indeed (1918 case1), could be linked to viral pneumonia and exhibited symptoms of acute pneumonia in the left lung combined with an acute form of bronchiolitis in the right lung, a pathological characteristic typical of a "primary viral pneumonia".

Focusing on case1, researchers performed control amplification of reverse-transcribed genetics of the nine gene fragments of the 1918 virus using the technique of polymerase chain reaction (PCR). They then carried out phylogenetic analyses based on the gene sequences to reconstruct the genealogical relationships between these elements. It was concluded that the genetic sequence of this strain was different from every other influenza strain, and that it was more closely related to strains found in birds than in mammals (Taubenberger et al. 1997). This partial analysis of the genetic map of human influenza was soon followed by a complete sequencing of the hemmagglutinin gene (HA) – a gene long believed to be "pivotal" in the pathogenicity of influenza A viruses (Webster and Rott 1987; see also Cox and Bender 1995). This gene codes for a protein located on the surface of the virus that plays a crucial role in allowing the virus to bind to host cells. If the virus is able to spread to another species this means it has somehow (through antigenic drift) acquired a new protein that enables it to bind on a different receptor. However, the team did not identify a mutation of the cleave site of the hemmagglutinin gene (Reid et al. 2001; Taubenberger et al. 1997).

Two years later, the team published another article on the "Origin and evolution of the 1918 'Spanish' influenza virus hemmagglutinin gene" (Reid, Fanning,

Hultin, and Taubenberger 1999). Johan Hultin, the pathologist who attempted to find traces of the influenza virus in Alaska in the early 1950s, was among the authors of the study. After reading the 1997 *Science* paper, Hultin wrote a letter to Taubenberger offering to return to Brevig Mission to look for samples of people who had died of the flu (Taubenberger 2003, 43). Against all odds, Hultin was successful. After he received the approval of Taubenberger he set out to Alaska for a second time and in August 1997 he found in situ frozen lung biopsies. Once in the village, he was granted permission from the council to dig the graveyard again; with the help of a few villagers and after 4 days work, he unearthed the body of a 30 year-old woman whom he called "Lucy". Opening up her chest he found two frozen lungs that he immediately sent to Taubenberger's laboratory in Washington, along with some tissues taken from three other frozen corpses (Berche 2012).

Reid et al. (1999) reported on the full sequence of the hemmagglutinin gene using RNA fragments from case1 discussed in the first article. They investigated three case histories to find evidence of influenza RNA. The first one was a 21 year-old man who died at Fort Jackson in South Carolina. Pathological records indicate he had pneumonia and influenza symptoms; he was admitted to the camp hospital on September 20th 1918 and died within 6 days. The autopsy records also show that his left lung suffered from an acute and fatal attack of pneumonia, whereas his right one showed acute bronchiolitis and alveolitis – a clear sign of influenza infection. No RNA was found in the left lung. However, the team performed a minute microscopic analysis on the paraffin-embedded tissue of the right lung and tissues tested positive for influenza RNA. The fragments of five genes were sequenced, amplified through PCR technique and then determined. The second case was also a male soldier, this one 30 years old and based at Camp Upton in the State of New York. He was admitted to hospital with pneumonia and died within 3 days on the 23rd of September 1918. Microscopic examination of his lungs by Taubenberger and his team revealed acute pulmonary oedema and acute bronchopneumonia. Formalin-fixed, paraffin-embedded samples of lung tissues tested positive for influenza RNA, the sequence of which was no longer than 150 nucleotides. The third case history was the one found by Hultin in Brevig Mission, Alaska.

Using the sequences of these three case histories, the Washington-based team worked out the genealogical relationships between them. Their analysis reasserted that the virus that caused the pandemic was avian in nature and that it entered human populations between 1900 and 1915, following the modification of the binding site on the HA protein. In 2005, Taubenberger and Tumpey published two separate articles in *Nature* and *Science*: the first provided the complete genomic sequence of the 1918 influenza virus and the second revealed the methods used to artificially reconstruct it. Yet, even before the complete genomic map of the virus was made available, it became unclear whether the genes of the influenza virus had indeed disclosed the causes of its exceptional virulence (see Taubenberger et al. 2001). Moreover, their argument of a likely avian origin of the virus was criticized.

A Missing Mutation and the Limits of Genomic Analyses

Efforts to sequence the virus that caused the 1918–19 influenza pandemic were motivated by the possibility of understanding the genetic origin and virulence of such an organism. While this work allowed for a more precise characterization of the hemagglutinin, neuraminidase, matrix, and nucleoprotein gene segments from a functional point of view, it is less clear, however, whether the first goal was achieved. In effect, the Washington team reported that a cleavage-site mutation on the hemmagglutinin gene that played a crucial role in the virulence of the Hong Kong pandemic in 1968 was not found in the strain obtained from the South Carolina case. Sequencing the specific cleavage site in the RNA of the virus obtained from the Brevig Mission case and New York case also confirmed that this mutation was absent. Inquiring into this mutation site (hemmagglutinin) – understood as a key determinant of virulence – was a central motivation of Taubenberger's work as it would have "offered an appealing explanation of the 1918's flu virulence" (Taubenberger 2003, 45). Yet Taubenberger was forced to recognize that "the 1918 strain (as confirmed by all three cases) does not possess a mutation at this site" (Ibid.; see also Reid et al. 1999; Stevens et al. 2004). In the light of this conclusion, virologist and influenza expert Robert Webster wrote that the secret of the Spanish influenza will "remain elusive". Webster commented that such "biological properties" [i.e. virulence] may "not be resolved" and suggested that the results of the sequencing project could only provide a partial explanation of this phenomenon. Indeed, for him "the entire gene sequence is unlikely to reveal the secret of the high pathogenicity of the 1918 Spanish virus" (Webster 1999, 1165). While Taubenberger's paper ends with some remarks about the complex, likely polygenic, nature of virulence determinants in a particular strain, it also concludes – contra Webster – with the hope that more sequencing would "shed additional light on the nature of the 1918 influenza virus" (Reid et al. 1999, 1656).

Another molecular explanation of the 1918–19 flu pandemic emerged in 1998 from another research team. Virologists Hideo Goto and Yoshihiro Kawaoka published a paper in the *Proceedings of the National Academy of Science* on a novel mechanism for the acquisition of virulence by human influenza A viruses. There, they argued that a change in another major protein – neuraminidase – able to increase the cleavage of HA could bring about higher levels of virulence. In fact, Goto and Kawaoka even suggested that a change in a single amino-acid sequestering plasminogen might facilitate the cleavage of NA. The authors were cautious, however, stating they "do not conclude that single mutation will convert nonplasminogen-binding NAs to efficient plasminogen binders, thus rendering the virus highly virulent" (1998, 10228). Yet, they acknowledged at the same time that it is "tempting to speculate that the 1918 pandemic strain [...] may have acquired its unprecedented virulence from the mechanism we describe" (Ibid). But such a change in amino-acid was also absent (or at least not observed) in the 1918 neuraminidase sequence (Taubenberger 2003, 45; 1988; see also Reid et al. 2000; Kawaoka and Watanabe 2011). Also, similar to Taubenberger, Goto and Kawaoka concluded with a plea for "further sequencing", in order "to address the issue of its [the 1918–19 pandemic] unprecedented virulence" (Goto and Kawaoka 1998, 10228).

In 2004, both Taubenberger and Tumpey acknowledged the lack of evidence provided by the molecular structure of the virus to explain its virulence:

> Sequence analysis of the 1918 influenza virus from fixed and frozen lung tissue has provided molecular characterization and phylogenetic analysis of this strain. The complete coding sequence of the 1918 nonstructural (NS), hemagglutinin (HA), neuraminidase (NA), and matrix (M) genes have been determined; however, the sequences of these genes did not reveal features that could account for its high virulence" (Tumpey et al. 2004; emphasis added).

And yet, despite evidence for an absence, there seems to be something particular about the structure of the HA protein that contributes to an enhanced level of virulence (Morange 2005). Indeed, using a mouse model, another team of molecular pathologists (Kobasa et al. 2004) showed that when the HA protein taken from the 1918 viral strain is inserted into mice it confers high pathogenicity and facilitates lung infections. For instance, infected mice show 39,000 times more virus particles after infection with the 1918 strain than with other viral strains like the Texas virus, and infected mice died after 6 days following infection with the 1918 strain, while all survived when infected with the Texas virus (von Bubnoff 2005, 794). The particular structure of the protein responsible for such pathological effect remains to be found, however, and it is unclear whether similar effects could hold true in humans as well.

Evolutionary Explanations in Emerging Diseases and Changes in Virulence

As we have described, Taubenberger's team provided the first molecular characterisation of the Spanish influenza organism based on the construction of phylogenetic trees of 9 of the 11 RNA-polymerase genes of the 1918–19 virus (Taubenberger 2005). The authors of this research project that spanned several years concluded that the virus did not originate from gene reshuffling (or reassortment) but rather that it jumped from birds to humans shortly before the onset of the pandemic. The virus was thus of avian origin. However, their interpretation of the similarity by descent, and thus of the genealogical relationships between the 1918 virus and today's avian viruses was disputed (Gibbs and Gibbs 2006; Antonovics et al. 2006).

As the current head of the Viral Pathogenesis and Evolution Section at the National Institute for Allergy and Infectious Diseases, Taubenberger's work is underpinned by evolutionary considerations. But what aspects of his work exactly are evolutionary or Darwinian? Philosopher Michael Ruse has long pointed out that the term "Darwinism" carries two broad meanings. It can be used firstly in a metaphysical sense to characterize change, development and transformation in the natural world. In this sense, the concept of Darwinism is older than Darwin himself. Another sense of Darwinism is important to acknowledge. In this second sense,

Darwinism is a scientific notion that emerges in the work of naturalist Charles Darwin and refers to the fact of evolution, the paths (phylogenies) of evolution, and the mechanism (natural selection) of evolution (Ruse 1992, 77). This distinction between path and mechanism maps on the more traditional distinction between patterns and processes in evolutionary biology mentioned above. The research of Taubenberger and Ewald – and more generally molecular pathology and evolutionary ecology – displays these two aspects of Darwinian theory. Arguably, both accept evolution as a "fact". However, the former is more interested in the "patterns" of evolution and uses evolutionary thinking to unravel the biological (including genetic) and adaptive processes that led to an increase in virulence. In contrast, Ewald focuses on the "process" of evolution – natural selection – as it occurred in various environments and populations of hosts and pathogens. As described above, Taubenberger's research focuses on precise and minute description of the small steps that allow viruses to infect more than one species; this work painstakingly tracks changes in nucleotides and charts the genealogical relationships between several strains of influenza. Ewald and Oxford, in contrast, take a broader view and ask why those mutations were selected, what were the selective pressures that drove them to be passed on and conserved in the gene pool, and especially, what is the role of the milieu, largely understood, in shaping virulence.

Though the centrality of the concept of natural selection is not really in dispute here, the ways in which Taubenberger and Ewald (and other evolutionary ecologists) understand these processes differs significantly on one important point: whereas the former describes the small incremental steps leading to the high, observable level of virulence, the latter looks for a plausible, eventually testable evolutionary scenario leading to the accumulation and conservation of these small, gradual changes. In other words, the second approach, the ecological one, seeks not only to describe organic changes leading to the formation of new viral strains, for example, but also attempts to give an account of the adaptive value of these transformations in the particular milieu in which the microorganisms lived, reproduced and eventually died. These two components of evolutionary theory – patterns and processes – are well known in the history of biology. Evolutionary ecologists nowadays might want to argue that Taubenberger is primarily interested in constructing and comparing distinct phylogenetic trees, no matter what the significance of their (evolutionary) relationship may be.[17] We think that the two aspects of evolutionary theory discussed here, however, reflect more broadly the existence of two distinct styles of scientific practices in biomedicine. The difficulty in addressing both aspects of the theory at the same time is indicative of a genuine tension between distinct explanatory strategies where knowledge claims are made according to different assumptions as to what counts as explanatory.

---

[17] Antonovics et al. (2006) critically wrote that, "the phylogenies described by Taubenberger et al. contradict their main conclusions and are presented without discussion of the evolutionary relationships they imply" (2006, E9).

A Note on the Plurality of the Scientific Styles of Practice

Confronted with the lack of evidence supporting a molecular explanation, and in the light of the limitations of the environmental-ecological account, one could have expected researchers to seek support in each other's work in order to complement their researches, and to move beyond the limitations of their own methodologies and research paradigms. Yet it is striking to note that Reid, Taubenberger et al. (1999), on the one hand, and Goto and Kawaoka (1998), on the other, reached a conclusion diametrically opposed to that of Webster and also Ewald: for the former, in order to explain better the influenza pandemic, more genomic sequencing is needed. Instead of considering other possible explanations of the exceptional virulence (i.e. ecological explanations) they persist in their attempt to provide a complete and satisfying explanation within a single explanatory framework.

At this point, a few remarks are in order. Firstly, and from a broad sociological point of view, this may just be a sign of our times: sequencing genetic material is an effective, and now rather inexpensive, way of obtaining prestigious research grants. Proposals in genomics, synthetic biology, and other cognate fields with a strong engineering approach to biology can highlight potential findings and even future applications, some of which are likely to be patentable and thus rapidly rentable from a financial point of view. In brief, promoting more sequencing is likely to provide additional research money. While this may be a reason why Taubenberger's team value more genetic sequencing other reasons of a more epistemological and historical nature must also be envisaged.

A second reason to consider has to do with what historians and philosophers of science have called a scientific "style of practice" (Keating and Cambrosio 2007). Derivative of Ian Hacking's concept of "styles of reasoning" – itself inspired by Alistair Crombie's "style of scientific thinking in the European tradition"– (Hacking 1992; Crombie 1994), the notion of "style" typically refers to the historical formation of distinctive practices and methodologies in science. Styles frame what counts as evidence, relevant questions to ask, truth-value, and sound explanation in distinct research and/or cultural contexts. Alongside the development of individual styles of practice one finds the emergence of new standards for measurements, objectivity, proof, and so on (Hacking 1992). Though styles are flexible they are not loose or relativist categories; they admit rules, systems of norms, stabilization techniques, and methods of justification. As they progressively become stabilised over time and entrenched within scientific activities, however, the very existence of styles of reasoning and their historical development become taken for granted. While the notion of style is often employed to analyse scientific controversies (Amsterdamska 2004; Fujimura and Chou 1994), it is interesting to note here that the two styles at play in the present case study have grown in relative ignorance of each other. Going back to the missing mutations, we can see that even though Taubenberger's programme did not provide the answers it sought it could not be halted hastily, especially after gathering immense publicity and funding. On the contrary, it is expected that these scientists, working within their style of practice, continue to do so until all possibilities of finding the key to the exceptional virulence have been looked at and examined in detail. From this point of view, their persistence in seeking a complete molecular explanation makes sense – even if, from a public health and biosecurity point of view, their research raises ethical concerns about the development of dual-use technologies (Rappert 2007). Moreover, the results obtained on the biology of influenza A viruses and the methods developed by Taubenberger and his team now enable worldwide researchers to better understand the molecular differences between various influenza strains.

What appears as a sign of determination in pursuing a research objective can also reflect a lack of communication between distinct scientific communities, the problems of interdisciplinary work, the self-containment of styles of scientific practice, and/or the resistance offered by epistemological obstacles. The current gap between ecological and molecular explanations, as it emerged in the present case, may be due to the fact that functional explanations such as those constructed in molecular biology tend to appear "self-sufficient", as historian and biologist Michel Morange recently put it (2011). This sort of epistemological obstacle means that, for many, there is no (obvious) need to complement molecular explanations with ecological considerations. To say that integration between mathematical modelling and molecular microbiological approaches has failed in this case would be going too far, however. Indeed, integration of ecological and molecular approaches of virulence evolution has not even been seriously attempted so far. Also, it would be misleading to suggest that molecular pathologists wholly ignore the environmental perspectives on virulence evolution and emerging diseases.[18] Yet when they do take them into account, the result does not necessarily amount to a better integration of data, theories, or methods but reveals, instead, the heights of disciplinary boundaries and the valuing of one style of practice over another. For example, in one of the last publications of Taubenberger and his colleagues at the National Institute of Health, the authors concluded that the diminution of severity of influenza pandemics over time "is surely due in part to advances in medicine and public health, but it may also reflect viral evolutionary choices that favor optimal transmissibility with minimal pathogenicity - *a virus that kills its host too fast or sends them to bed is not optimally transmissible*" (Morens et al. 2009, 229; emphasis added). In other words, the biological interests of the virus will best be served by evolving lower virulence over time in order to facilitate transmission to new hosts, an explanation that rests on the conventional wisdom rejected by most evolutionary ecologists who advocate the theoretical trade-off model but that is still defended by some microbiologists. This may come as a surprise given that Anthony Fauci, on of the authors of the paper, and current head of the National Institute for Allergy and Infectious Diseases, has long criticized this view (see Fauci 2005). It shows, however, that branches of sciences in which the same problem is addressed, through distinct methodologies, can be surprisingly disconnected and separated by epistemic gaps, professional or institutional barriers. In other words, integration is no easy goal to achieve.

---

[18] For instance, John Oxford collaborated on a paper with Reid, Taubenberger and several others in 2003 (Reid, Janczewski, Lourens et al. 2003).

Problems within the molecular style of practice, however, are not only epistemological but also ethical and social. The publication of the whole sequence of the 1918–19 strain in 2005 sparked lively debates among scientists and the public as it raised concerns as to whether it was safe to publish the methodology used to resurrect the pathogen (Rappert 2007; Selgelid 2005). What if someone with nefarious intentions reconstructs the virus? How likely is it that this genetic information be used for harmful purposes? What if, by accident or not, the virus escapes into the environment? For some, like biologist Richard H. Ebright from Rutgers University, "there is a risk, verging on inevitability, of accidental release of the virus" but "there is also a risk of deliberate release of the virus".[19] Yet others argued that the work of Taubenberger and Tumpey was entirely legitimate and could be applied to other areas and problems in virology such as the H5N1 pandemic and "could have an immediate impact by helping scientists focus on detecting changes in the evolving H5N1 virus that might make widespread transmission among humans more likely".[20]

The case of the Spanish influenza pandemic is today a classical example of a technology that has the potential for "dual-use" research (i.e. it could help to understand the disease and fight it, but it also could be used to disseminate it further in a population). A recent case of potential dual-use consequences in influenza research involving a group of researchers led by Ron Fouchier in the Netherlands and by Yoshihiro Kawaoka, from Madison, in the U.S, led in January 2012 to a 60 day suspension of research on influenza and virulent diseases, following consensus to delay publication.[21] Both teams had submitted a paper, to *Science* and *Nature* respectively, describing the methodology employed to artificially render an H5N1 influenza strain transmissible between ferrets (which is, arguably, a reliable indicator of possible transmission to humans) due to a mutation on the hemmagglutinin protein. Both studies have now been published (Imai et al. 2012; Herfst et al. 2012). As an aside, it is interesting to see that these two studies, although undoubtedly driven by molecular biology questions, were based on classical approaches in evolutionary ecology known as serial passage experiments (Ebert 1998).

In the early 1990s, the Institute of Medicine's report on *Microbial Threats* (Lederberg, Shope, and Oaks 1992) and Stephen Morse's *Emerging Viruses* (Morse 1993) have emphasized how emerging infectious diseases are posing a renewed threat to public health that needs to be addressed on a global scale, from the combined perspective of ecological and molecular approaches. The concept of emerging diseases has helped focus international efforts to contain infectious diseases within well-defined geographical and temporal limitations. With the (re)creation of the 1918–19 influenza strain and others (e.g. *Yersinia pestis*, polio virus, H5N1), a different form of biological threat arises and it requires different political, institutional, and legal response mechanisms. Indeed, while the threat of emerging infections was mostly perceived as coming from outside Northern-hemisphere countries,

---

[19] From New York Times article by Gina Kolate (2005) which can be accessed at http://www.nytimes.com/2005/10/06/health/06flu.html?pagewanted=all

[20] This statement was jointly issued by Fauci and Gerberding. See Kolate (2005).

[21] http://www.who.int/mediacentre/news/releases/2012/h5n1_research_20120217/en/index.html.

it now appears to be growing from within the heartland of Western countries itself. Instead of stressing possible disease invasions in previously unexposed countries (or with only low incidence of a particular disease), recently developed technologies in synthetic biology and genomics have opened-up the possibility to artificially create new diseases, or to resurrect old ones such as the plague, the influenza strain responsible for the 1918 pandemic, and the polio virus (Bos et al. 2011; Taubenberger 2005; Tumpey et al. 2005; Celo et al. 2002; Rosengard et al. 2002). On the other hand, it might be argued that new variants appear constantly and the risk of a laboratory accident might be comparable to what happens naturally in the field. Moreover, while dual-use is a characteristic of most life sciences nowadays (Atlas 2009), only a small number of experiments and experimental practices are, overall, seen as posing real threats to public health and global security (for a recent analysis see Aucouturier 2011; Morens et al. 2012). Finally, it is worth noting that dual-use technologies – like scientific research more generally – are often characterized by unexpected findings such as, for instance, the accidental discovery that a modified virus injected into mice was lethal to otherwise vaccinated animals (Jackson et al. 2001). As is often the case in science, the experimental system designed to answer certain questions opens-up theoretical and practical possibilities that could sometimes not be envisaged at the outset (Rheinberger 1997). If unpredictability and unforeseen results are truly the essence of scientific research, dual-use technologies are then an unavoidable trade-off to deal with, a point that reinforces the need to develop appropriate governance responses to biomedical research programmes on pathogens and potentially pathogenic organisms (Méthot 2014). More generally, those new research avenues underline the need for the development of a "culture of responsibility" (NSABB 2011) in the life sciences, that is, a new ethos to address and balance questions of biosecurity and risk with scientific autonomy and progress, among others.

Concluding Remarks

The two most important glycoproteins allowing influenza viruses to invade host tissues – hemmagglutinin (HA) and neuraminidase (NA) – were significant molecular determinants of the virulence of influenza pandemics in 1957 and 1968, and can yield potentially pathogenic effects when inserted into some animal models. Considerable explanatory power was placed on these special proteins that seemed to provide a first-hand, adequate, simple and certainly elegant mechanism to account for the exceptional virulence of the 1918 pandemic. Indeed, the "most popular theory" was that the 1918 virus had "unique pathogenic properties, most likely encoded within the hemagglutinin protein" (Holmes 2004). Identifying a molecular and genetic basis of virulence could not only provide a window into the most devastating epidemic of modern times, but could also help to prevent and predict those to come. Overall, the remarkable technological success – i.e. the retrieving and sequencing of the 1918 avian virus – promised nothing less than to unlock one of the oldest and well-kept secrets in the whole of medical history. However, after sequencing the genome of the 1918 viral strain that killed perhaps up to 40 million people according to the WHO estimates, both factors were found to be lacking in the killer strain.

One might wonder about the extent to which it is possible to generalize from the example of the Spanish influenza pandemic to other cases. Researchers on ancient pathogens using high throughput technologies have recently claimed to identify the causal organism of the Black Death (*Yersinia pestis*) in the fourteenth century and the sources of its virulence in the form of a single plasmid (Schuenemann et al. 2001). However, determining why an organism is pathogenic or what makes it a pathogen is not straightforward and is rarely based on a specific structural characteristic alone (Méthot 2012c). As microbiologist Charles Nicolle once said (1930), virulence is the expression of a "mosaic of powers" resulting from a constellation of factors that are irreducible to any particular structure and must be understood against a broad biological and even historical background. It is interesting to note, therefore, that the same team went on to revise its position in a subsequent article by pointing out the inherent limitations of molecular-oriented explanation and, furthermore, emphasized the need to widen the explanation and integrate ecological factors as well. They write:

> Regardless, although no extant *Y. pestis* strain possesses the same genetic profile as our ancient organism, our data suggest that few changes in known virulence-associated genes have accrued in the organism's 660 years of evolution as a human pathogen, further suggesting that its perceived increased virulence in history may not be due to novel fixed point mutations detectable via the analytical approach described here. At our current resolution, we posit that molecular changes in pathogens are but one component of a *constellation of factors* contributing to changing infectious disease prevalence and severity, where *genetics of the host population, climate, vector dynamics, social conditions and synergistic interactions with concurrent diseases should be foremost in discussions of population susceptibility to infectious disease and host–pathogen relationships* with reference to *Y. pestis* infections (Bos et al. 2011; emphasis added).

In sum, the study of Bos et al. (2011) did not reveal any significant genetic or evolutionary change in 600 years that could explain the virulence of plague in the fourteenth century. As a consequence, they argue that a molecular approach only provides an incomplete picture when applied in isolation, and that a complementary ecological perspective is needed. More precisely, the more recent study emphasizes that a full understanding of the evolution of virulence requires a multi-dimensional framework that encompasses host resistance, ecological factors, and the interactions between the different diseases occuring in a well-defined geographical area over a specific time period. To go beyond the limitations of analytical approaches that investigate one disease at a time, a synthetic and global approach is necessary in order to understand more broadly the evolution of emerging diseases that compose the past, present, and future of any "pathocenosis" (Grmek 1969).

To conclude, our analyses of the case of the Spanish influenza pandemic show that there is an irreducible tension in studying the phenomenon of virulence comparatively across the sciences: one can either look for determinants of pathogenic power analytically, that is from within microorganisms, or synthetically at the level of host-pathogen interactions in a given ecological environment. Both approaches face limitations and neither appears to be sufficient to account for a complex phenomenon like the 1918–19 influenza pandemic. Besides, each level of analysis has its own ways of characterizing the nature and causes of virulence (and pathogenicity), and both lead to the development of different measures to prevent or to treat emerging infectious diseases. While ecological approaches contribute to the establishment of national and international programmes intended to increase detection and surveillance in emerging infections on a global scale, monitoring changes in virulence to prevent pandemics worldwide, molecular approaches facilitate the development of biomedical tools, such as vaccines or antibiotics, to fight infectious diseases either by reducing their pathogenic power, and/or by enhancing individual and group (or herd) immunity. Yet both perspectives can also work together: molecular phylogenies can provide evidence regarding the likely origins of future pandemics and help to channel the attention of a detection network onto particularly sensitive sites (see Morens et al. 2012). A better integration of ecological and molecular approaches would thus benefit public health medicine by providing stronger theoretical approaches and empirical mechanistic models to understand, manage and perhaps even predict future influenza pandemics.

Acknowledgments Earlier versions of this article were presented by Pierre-Olivier Méthot at Bristol University as part of the Seminar Series in the Philosophy of Medicine in March 2011 and during the Consortium in the History and Philosophy of Biology at the IHPST, in Paris, in June 2011. Financial support from the SSHRC is gratefully acknowledged (no.752-2007-1257).
Samuel Alizon is funded by an ATIP-avenir grant from CNRS and INSERM.

References

Alizon S (2014) Evolution of disease virulence. In: Losos J (ed) Oxford bibliographies in evolutionary biology. Oxford University Press, New York, 01/13/2014, doi:10.1093/OBO/9780199941728-0018

Alizon S, Michalakis Y (2011) The transmission-virulence trade-off and superinfection: a comment to Smith. Evolution 65:3633–3638

Alizon S, van Baalen M (2005) Emergence of a convex trade-off between transmission and virulence. Am Nat 165:E155–E167

Alizon S, Hurford A, Mideo N, Van Baalen M (2009) Virulence model and the trade-off hypothesis: history, current state of affairs and the future. J Evol Biol 22:245–259

Amsterdamska O (2004) Achieving disbelief: thought styles, microbial variation, and American and British epidemiology, 1900–1940. Stud Hist Philos Biol Biomed Sci 35(3):483–507

Anderson RM, May RM (1982) Coevolution of hosts and parasites. Parasitology 85:411–426

Anderson RM, May RM (1991) Infectious diseases of humans: dynamics and control. Oxford University Press, Oxford

Antia R, Regoes RR, Koella JC, Bergstrom CT (2003) The role of evolution in the emergence of infectious diseases. Nature 426:658–661

Antonovics J, Hood ME, Baker CH (2006) Was the 1918 flu avian in origin? Nature 440:E9

Atlas R (2009) Responsible conduct by life scientists in an age of terrorism. Sci Eng Ethics 15(3):293–301

Aucouturier E (2011) La guerre biologique: perspective historique et critique. PhD thesis, Université Paris1 (Panthéon-Sorbonne)

Ball GH (1943) Parasitism and evolution. Am Nat 77(771):345–367

Barry JM (2004a) The great influenza: the epic story of the deadliest plague in history. New York, Viking Books

Barry JM (2004b) The site of the origin of the 1918 influenza pandemic and its public health implications. J Transl Med 2(3)

Berche P (2012) Faut-il encore avoir peur de la grippe? Histoire des pandémies. Odile Jacob, Paris

Bergstrom C (2008) An evolutionary biologist considers the virulence of emerging infectious diseases. Nature Journal Club. 19th May, p 261

Beveridge WIB (1992) The chronicle of influenza pandemics. Hist Philos Life Sci 13:223–235

Boots M, Best A, Miller MR, White A (2009) The role of ecological feedbacks in the evolution of host defence: what does theory tell us? Philos Trans R Soc Lond B Biol Sci 364:27–36

Boots M, Sasaki A (1999) 'Small worlds' and the evolution of virulence: infection occurs locally not at a distance. Proc R Soc Lond B 266:1933–1938

Bos KB, Schuenemann VJ, Golding GB, Burbano HA et al (2011) A draft genome of Yersinia pestis from victims of the Black Death. Nature 478:506–510

Bouchard F (2004) Evolution, fitness, and the struggle for persistence. PhD. thesis, Duke University

Bresalier M (2013) A short history of flu. Bloomsbury Academic, London

Bull JJ (1994) Virulence. Evolution 48(5):1423–1437

Bull JJ, Levin BR (1994) Parasites on the move. Science 265:1469–1470

Burnet FM (1946) Virus as organism. Evolutionary and ecological aspects of some human virus diseases. Harvard University Press, Cambridge, MA

Burnet FM (1953) The future of medical research. The Lancet i:103–108

Burnet FM (1960) Chairman's opening remarks. In: Wolstenholme GE, O'Connor CM (eds) Virus, virulence and pathogenicity. Churchill, London, pp 1–2

Burnet FM, Clark E (1942) Influenza, Monographs from the Walter and Eliza Hall Institute of Research in Pathology and Medicine. Macmillan Company, Melbourne

Bush RM (2007) Influenza evolution. In: Tybayrenc M (ed) Encyclopedic guide to infectious disease research: modern methodologies. Wiley, Hoboken, pp 199–214

Castillo-Salgado C (2010) Trends and directions of global public health surveillance. Epidemiol Rev 32:93–109

Celo J, Paul AV, Wimmer E (2002) Chemical synthesis of poliovirus cDNA: generation of infectious virus in the absence of a natural template. Science express. Report. www.sciencexpress.org

Chapleau F, Johansen PH, Williamson M (1988) The distinction between pattern and process in evolutionary biology: the use and abuse of the term 'strategy'. Oikos 53(1):136–138

Cockburn TA (1963) The evolution and eradication of infectious diseases. Johns Hopkins Press, Baltimore

Cox N, Bender CA (1995) The molecular epidemiology of influenza viruses. Virology 6:359–370

Crombie AC (1994) Styles of scientific thinking in the European tradition: the history of argumentation and explanation especially in the mathematical and biomedical sciences and arts. Duckworth, London

Crosby AW (1989) America's forgotten pandemic. The Influenza of 1918. Cambridge University Press, Cambridge

Day T, Proulx SR (2004) A general theory for evolutionary dynamics of virulence. Am Nat 163(4):E41–E63

Diekmann O, Heesterbeek JAP (2000) Mathematical epidemiology of infectious diseases. Model building, analysis, and interpretation. Wiley, New York

Ebert D (1998) Experimental evolution of parasites. Science 282:1432–1435. doi:10.1126/science.282.5393.1432

Ewald PW (1983) Host-parasite relations, vectors and the evolution of disease severity. Annu Rev Ecol Syst 14:465–485

Ewald PW (1988) Cultural vectors, virulence, and the emergence of evolutionary epidemiology. Oxf Surv Evol Biol 5:215–245

Ewald PW (1991) Transmission modes and the evolution of virulence, with special reference to cholera, influenza, and AIDS. Hum Nat 2(1):1–11

Ewald PW (1994) Evolution of infectious disease. Oxford University Press, Oxford

Ewald PW (1996) Guarding against the most dangerous emerging pathogens: insights from evolutionary biology. Emerg Infect Dis 2:245–257

Fantini B (1993) Les organisations sanitaires internationales face à l'émergence des maladies infectieuses nouvelles. Hist Philos Life Sci 15:435–457

Farmer P (1996) Social inequalities and emerging diseases. Emerg Infect Dis 2(4):259–269

Fauci AS (2000) Infectious diseases: consideration for 21st century. Clin Infect Dis 32:675–685

Fauci AS (2005) Emerging and re-emerging infectious diseases: the perpetual challenge, Robert H. Ebert memorial lecture. Milbank Memorial Fund, New York, pp iv–18

Fenner F, Fantini B (1999) Biological control of vertebrate pests. The history of myxomatosis. An experiment in evolution. CABI, New York

Fleck L (1979) Genesis and development of a scientific fact. Chicago University Press, Chicago [First published 1935]

Frank SA, Schmid-Hempel P (2008) Mechanisms of pathogenesis and the evolution of parasite virulence. J Evol Biol 21:396–404

Fraser C, Hollingsworth TD, Chapman R et al (2007) Variation in HIV-1 set-point viral load: epidemiological analysis and an evolutionary hypothesis. Proc Natl Acad Sci U S A 104(44): 17441–17446

Fraser C, Cummings DAT, Klinkenberg D et al (2011) Influenza transmission in households during the 1918 pandemic. Am J Epidemiol 174(5):505–514

Friesen TL, Stukenbrock EH, Lui Z et al (2006) Emergence of a new disease as a result of interspecific virulence gene transfer. Nat Genet 38(8):953–956

Fujimura JH, Chou DY (1994) Dissent in science: styles of scientific practice and the controversy over the cause of AIDS. Soc Sci Med 38(8):1017–1036

Garrett L (1994) The coming plague: newly emerging diseases in a world out of balance. Farrar, Straus and Giroux, New York

Gibbs MJ, Gibbs AJ (2006) Was the 1918 pandemic caused by a bird flu? Nature 440:E8

Godfrey-Smith PJ (2006) The strategy of model-based science. Biol Philos 21:725–740

Goto H, Kawaoka Y (1998) A novel mechanism for the acquisition of virulence by a human influenza A virus. Proc Natl Acad Sci U S A 95:10224–10228

Graham AL, Allen JE, Read A (2005) Evolutionary causes and consequences of immunopathology. Annu Rev Ecol Evol Syst 36:373–397

Grmek MD (1969) Préliminaires d'une étude historique des maladies. Annales ESC 24:1473–1483

Grmek MD (1993) Le concept de maladie émergente. Hist Philos Life Sci 15:281–296

Groisman EA, Ochman H (1996) Pathogenicity islands: bacterial evolution in quantum leaps. Cell 87:791–794

Hacking I (1992) Styles for historians and philosophers. Stud Hist Philos Sci 23(1):1–20

Herfst S, Schrauwen EJA, Linster M, Chutinimitkul S, de Wit E, Munster VJ, Sorrell EM, Bestebroer TM, Burke DF, Smith DJ, Rimmelzwaan GF, Osterhaus ADME, Fouchier RAM (2012) Airborne transmission of influenza A/H5N1 virus between ferrets. Science 336(6088): 1534–1541. doi:10.1126/science.1213362

Holmes EC (2004) 1918 and all that. Science 303(5665):1787–1788

Honigsbaum MA (2013) History of the great influenza pandemic. Death, panic, and hysteria 1830–1920. I.B. Tauris Academic, London

Imai M, Watanabe T, Hatta M, Das SC, Ozawa M, Shinya K, Gongxun Zhong, Hanson A, Katsura H, Watanabe S, Chengjun Li, Kawakami E, Yamada S, Kiso M, Suzuki Y, Maher EA, Neumann G, Kawaoka Y (2012) Experimental adaptation of an influenza H5 HA confers respiratory droplet transmission to a reassortant H5 HA/H1N1 virus in ferrets. Nature. doi:10.1038/nature10831

Jackson RJ, Ramsay AJ, Christensen CD, Beaton S, Hall DF, Ramshaw IA (2001) Expression of mouse interleukin-4 by a recombinant ectromelia virus suppresses cytolytic lymphocyte responses and overcomes genetic resistance to mousepox. J Virol 75:1205–1210

Johnson N (2006) Britain and the 1918–19 Influenza pandemic. A Dark Epilogue. Routledge, New York

Johnson NPAS, Mueller J (2002) Updating the account: global mortality of the 1918–1920 « Spanish » Influenza pandemic. Bull Hist Med 76(1):105–115

Jones EK, Patel NG, Levy MA, Storeygard A, Balk D, Gittleman JL, Daszak P (2008) Global trends in emerging infectious diseases. Nature 451:990–994

Kawaoka Y, Watanabe T (2011) Pathogenesis of the 1918 pandemic influenza virus. PLOS Pathogens 7(1):1–4

Keating P, Cambrosio A (2007) Cancer clinical trials: the emergence and development of a new style of practice. Bull Hist Med 81(1):197–223

Kilbourne ED (ed) (1960) The severity of influenza as a reciprocal of host susceptibility. In virus virulence and pathogenicity. Churchill, London

Kilbourne ED (1977) Influenza pandemics in perspective. JAMA 237(12):1225–1227

King N (2004) The scale politics of emerging diseases. Osiris 19:62–76

Kobasa D, Takada A, Shinya K et al (2004) Enhanced virulence of influenza A viruses with the haemagglutinin of the 1918 pandemic virus. Nature 431:703–707

Kolate G (2005) http://www.nytimes.com/2005/10/06/health/06flu.html?pagewanted=all

Krause RM (1998) Introduction. In: Krause RM (ed) Emerging infections. Biomedical research report. Academic Press, London

Laland KN, Sterelny K, Odling-Smee J, Hoppitt W, Uller T (2011) Cause and effect in biology revisited: Is Mayr's proximate-ultimate dichotomy still useful? Science 334(6062):1512–1516

Langford C (2005) Did the 1918–19 influenza pandemic originate in China? Popul Dev Rev 31(3):473–505

Lederberg J (1993) Viruses and humankind: intracellular symbiosis and evolutionary competition. In: Morse SS (ed) Emerging viruses. Oxford University Press, New York/Oxford, pp 3–9

Lederberg J, Shope RE, Oaks SC (1992) Emerging infections: microbial threats to health in the United States. National Academy Press, Washington, DC

Levin BR (1996) The evolution and maintenance of virulence in microparasites. Emerg Infect Dis 2(2):93–102

Levin BR, Edén CS (1990) Selection and evolution of virulence in bacteria: an ecumenical excursion and modest suggestion. Parasitology 100:103–115

Levin SA, Pimentel D (1981) Selection for intermediate rates of increase in parasite-host systems. Am Nat 111:3–24

Levins R (1994) The challenge of new disease. In: Wilson ME, Levins R, Spielman A (eds) Disease in evolution. Global changes and emergence of infectious diseases. The New York Academy of Sciences, New York, pp xvii–xix

Lipsitch M, Moxon RE (1997) Virulence and transmissibility of pathogens: what is the relationship? Trends Microbiol 5(10):31–37

Maurelli AT (2006) Black holes, antivirulence genes, and gene inactivation in the evolution of bacterial pathogens. Microbiol Rev 267:1–8

May RM, Anderson RM (1983) Parasite host coevolution. In: Futuyama DJ, Slatkin M (eds) Coevolution. Sinauer, Sunderland, pp 186–206

Mayr E (1961) Cause and effect in biology: kinds of causes, predictability, and teleology are viewed by a practicing biologist. Science 134:1501–1506

McNeill WH (1976) Plagues and people. Basil Blackwell, Oxford

Mendelsohn JA (2002) 'Like all that lives': biology, medicine and bacteria in the age of Pasteur and Koch. Hist Philos Life Sci 24(1):3–36

Méthot PO (2012a) Why do parasites harm their hosts? On the origin and legacy of Theobald Smith's 'law of declining virulence' – 1900–1980. Hist Philos Life Sci 34:561–601

Méthot P-O (2012b) Historical epistemology of the concept of virulence: molecular, ecological, and evolutionary aspects of emerging infectious diseases in the 19th and 20th century. PhD thesis, University of Exeter & Université Paris 1 (Panthéon-Sorbonne)

Méthot PO (2012c) Understanding pathogens in the era of next-generation sequencing. J Infect Dev Ctries 6(9):689–691

Méthot P-O. (2014) Science and science policy: regulating 'select agents' in the age of synthetic biology. Perspect Sci 23(4) (Forthcoming)

Méthot PO, Alizon S (forthcoming) What is a pathogen? Towards a process view of host-parasite interactions. Virulence

Mills CE, Robins JM, Lipstich M (2004) Transmissibility of the 1918 pandemic influenza. Nature 432:904–906

Mitchell S (2008) Unsimple truths: science, complexity, and policy. University of Chicago Press, Chicago

Morange M (2005) Les secrets du vivant. Contre la pensée unique en biologie. La Découverte, Paris

Morange M (2011) What will result from the interaction between functional and evolutionary biology? Stud Hist Philos Biol Biomed Sci 42:69–74

Morens DM, Taubenberger JK (2009) Understanding influenza backward. JAMA 302(6):679–680

Morens DM, Folkers GK, Fauci AS (2004) The challenge of emerging and re-emerging infectious diseases. Nature 430(6996):242–249

Morens DM, Taubenberger JK, Fauci AS (2009) The persistent legacy of the 1918 influenza virus. N Engl J Med 361(3):225–229

Morens DM, Subbarao K, Taubenberger JK (2012) Engineering H5N1 avian influenza viruses to study human adaptation. Nature 486:335–340

Morse SS (1991) Emerging viruses: defining the rules for viral traffic. Perspect Biol Med 34(3):387–409

Morse SS (1993) Examining the origins of emerging viruses. In: Morse SS (ed) Emerging viruses. Oxford University Press, New York/Oxford

Morse SS (1995) Factors in the emergence of infectious diseases. Emerg Infect Dis 1(1):7–15

Morse SS (2007) Pandemic influenza: studying the lessons of history. Proc Natl Acad Sci 104(18):7313–7314

National Science Advisory Board for Security (NSABB) (2011) Guidance for Enhancing personal reliability and strengthening the culture of responsibility. http://osp.od.nih.gov/sites/default/files/resources/CRWG_Report_final.pdf

Nicolle C (1930) Naissance, vie et mort des maladies infectieuses. Félix Alcan, Paris

Noymer A (2010) Epidemics and time. Influenza and tuberculosis during and after the 1918–1919 pandemic. In: Herring DA, Swedlund AC (eds) Plagues and epidemics. Infected spaces past and present. Berg, Oxford/New York

O'Malley MA (2013) When integration fails: prokaryote phylogeny and the tree of life. Stud Hist Philos Biol Biomed Sci 44:551–562

O'Malley MA, Soyer OS (2012) The roles of integration in molecular systems biology. Stud Hist Philos Biol Biomed Sci 43(1):58–68

Olson DR, Simonsen L, Edelson PJ, Morse SS (2005) Epidemiological evidence of an early wave of the 1918 influenza pandemic in New York City. Proc Natl Acad Sci U S A 102(31):11059–11063

Oxford JS (2001) The so-called great Spanish influenza pandemic of 1918 may have originated in France in 1916. Philos Trans R Soc Lond B Biol Sci 356:1857–1859

Oxford JS, Sefton A, Jackson R, Innes W, Daniels RS, Johnson N (2002) World War I may have allowed the emergence of "Spanish" influenza. Lancet Infect Dis 2:111–114

Oxford JS, Lambkin R, Sefton A, Daniels R, Elliot A, Brown R, Gill D (2005) A hypothesis: the conjunction of soldiers, gas, pigs, ducks, geese and horses in Northern France during the Great War provided the conditions for the emergence of the "Spanish" influenza pandemic of 1918–1919. Vaccines 23:940–945

Palumbi SR (2001) Human as the world's greatest evolutionary force. Science 293:1786–1790

Pasteur L, Chamberland CE, Roux E (1994 [1881]) De l'atténuation des virus et de leur retour à la virulence. In: Pichot A (ed) Louis Pasteur. Écrits scientifiques et médicaux. Flammarion, Coll. Champs Sciences, Paris, pp 251–258

Phillips H, Killingray D (eds) (2003) The Spanish influenza pandemic of 1918–19. New perspectives, Studies in the social history of medicine. Routledge, London/New York

Poulin R, Combes C (1999) The concept of virulence. Parasitol Today 15(12):474–475

Raberg L, Graham AL, Read AF (2009) Decomposing health: tolerance and resistance to parasites in animals. Philos Trans R Soc 364:37–49

Rappert B (2007) Biotechnology, security and the search for limits. Palgrave Macmillan, New York

Read AF (1994) The evolution of virulence. Trends Microbiol 2(3):73–76

Reid AH et al (2003) The origin of the 1918 pandemic influenza virus: a continuing enigma. J Gen Virol 84:2285–2292

Reid AH, Fanning TG, Hultin JV, Taubenberger JK (1999) Origin and evolution of the 1918 "Spanish" influenza virus hemagglutinin gene. Proc Natl Acad Sci U S A 96:1651–1656

Reid AH, Fanning TG, Janczewski TA, Taubenberger JK (2000) Characterisation of the 1918 "Spanish" Influenza virus neuraminidase gene. Proc Natl Acad Sci U S A 97(12):6785–6790

Reid AH, Janczewski TA, Lourens RM, Elliot AJ, Daniels RS, Berry CL, Oxford JS, Taubenberger JK (2003) Influenza pandemic caused by highly conserved viruses with two receptor-binding variants. Emerg Infect Dis 9(10):1249–1253

Reid AH, Taubenberger JK, Fanning TG (2001) The 1918 Spanish influenza: integrating history and biology. Microbes Infect 3:81–87

Rheinberger HJ (1997) Towards a history of epistemic things. Synthesizing proteins in the test tube. Stanford University Press, Stanford

Rosengard AM, Liu Y, Nie Z, Jimenez R (2002) Variola virus immune evasion design: expression of a highly efficient inhibitor of human complement. Proc Natl Acad Sci U S A 99(13):808–813

Ruse M (1992) Darwinism. In: Keller, Loyds (eds) Keywords in evolutionary biology. Harvard University Press, Cambridge, MA, pp 74–80

Schuenemann VJ et al (2001) Targeted enrichment of ancient pathogens yielding the pPCP1 plasmid of Yersinia pestis from victims of the Black Death. Proc Natl Acad Sci U S A 108(38):E74–E752

Selgelid MJ (2005) Ethics and infectious disease. Bioethics 19(3):272–289

Shaner G, Stromberg EL, Lacy GH, Barker KR, Pirone TP (1992) Nomenclature and concepts of pathogenicity and virulence. Annu Rev Pythopathol 30:47–66

Shapiro-Ilan DI, Fuxa JR, Lacey LA, Onstad DW, Kaya HK (2005) Definitions of pathogenicity and virulence in invertebrate pathology. J Invertebr Pathol 88:1–7

Smith W, Andrewes CH, Laidlaw PP (1933) A virus obtained from influenza patients. Lancet 8:66–68

Snowden FM (2008) Emerging and reemerging diseases: a historical perspective. Immunol Rev 225:9–26

Stevens J, Corper AL, Basler CF, Taubenberger JK, Palese P, Wilson IA (2004) Structure of the uncleaved human H1 hemagglutinin from the extinct 1918 influenza virus. Science 303:1866–1869

Stuart-Harris CH (1953) Influenza. Arnold, London

Taubenberger JF (2005) Chasing the elusive 1918 virus: preparing for the future by examining the past. In: The threat of pandemic influenza: are we ready? Workshop summary, National Academy of Sciences, Washington, DC, pp 69–89

Taubenberger JF (2006) The origin and virulence of the 1918 "Spanish" influenza virus. Proc Am Philos Soc 150:86–112

Taubenberger JK (2003) Genetic characterization of the 1918 'Spanish' influenza pandemic. In: Philips H, Killingray D (eds) The Spanish influenza pandemic of 1918–19. New perspectives. 2003. Studies in the social history of medicine. Routledge, London/New York, pp 39–46

Taubenberger JK, Morens DM (2010) Influenza: the once and future pandemic. Pub Health Rep 125(Supplement 3):16–26

Taubenberger JK, Reid AH, Kraft AE, Bijwaard KE, Fanning TG (1997) Initial genetic characterisation of the 1918 "Spanish" influenza virus. Science 275:1793–1796

Taubenberger JK, Reid AH, Fanning TG (2000) The 1918 influenza virus: a killer comes into view. Virology 274:241–245

Taubenberger JK, Reid AH, Janczewski A, Fanning TG (2001) Integrating historical, clinical and molecular genetic data in order to explain the origin and virulence of the 1918 Spanish influenza virus. Philos Trans R Soc Lond 356:1829–1839

Thomas L (1974) The lives of a cell: notes of biology watcher. Penguin Books, New York

Thomas SR, Elkinton JS (2004) Pathogenicity and virulence. J Invertebr Pathol 85:146–151

Thompson WW, Shay DK, Weintraub E et al (2003) Mortality associated with influenza and respiratory syncytial virus in the United States. JAMA 289(2):179–186

Tumpey T, Garcia-Sastre A, Taubenberger JK, Palese P, Swayne DE, Basler C (2004) Pathogenicity and immunogenicity of influenza viruses with genes from the 1918 pandemic virus. Proc Natl Acad Sci U S A 101:3166–3171

Tumpey TM, Basler CF, Aguilar PV, Zeng H, Solorzano A, Swayne DE, Cox NJ, Katz JM, Taubenberger JK, Palese P, Garcia-Sastre A (2005) Characterization of the reconstructed 1918 Spanish influenza pandemic virus. Science 310(77):77–80

Van Helvoort T (1993) A bacteriological paradigm in influenza research in the first half of the twentieth century. Hist Philos Life Sci 15:3–21

Van Helvoort T (1994) The construction of bacteriophage as bacterial virus: linking endogenous and exogenous thought styles. J Hist Biol 27(1):91–139

Von Bubnoff A (2005) The 1918 flu virus resurrected. Nature. Special report: 794–795

Webby RJ, Webster RG (2003) Are we ready for pandemic influenza? Science 302:1519–1521

Webster RG (1993) Influenza. In: Morse SM (ed) Emerging viruses. Oxford University Press, New York/Oxford, pp 37–45

Webster RG (1999) 1918 Spanish influenza: the secrets remain elusive. Proc Natl Acad Sci U S A 96:1164–1166

Webster RG, Kawaoka Y (1994) Influenza – an emerging and re-emerging disease. Virology 5:103–111

Webster RG, Rott R (1987) Influenza virus A pathogenicity: the pivotal role of hemagglutinin. Cell 50:665–666

Webster RG, Bean WJ, Gorman OT, Chambers TM, Kawaoka Y (1992) Evolution and ecology of influenza A viruses. Microbiol Rev 56(1):152–179

Weir L, Mykhalovski E (2010) Global public health vigilance. Creating a world on alert. Routledge, New York/London

Witte W (2003) The plague that was not allowed to happen. German medicine and the influenza epidemic of 1918–19 in Baden. In: Philips H, Killingray D (eds) The Spanish influenza pandemic of 1918–19. New perspectives. 2003. Studies in the social history of medicine. Routledge, London/New York, 49–57

Wolfe ND, Dunavan CP, Diamond J (2007) Origins of major human infectious diseases. Nature 447:279–283

Power, Knowledge, and Laughter: Forensic Psychiatry and the Misuse of the *DSM*

Patrick Singy

Abstract This essay examines the relation between the DSM and forensic psychiatry. Psychiatrists, lawyers and philosophers often assume that the forensic legitimacy of the DSM hinges on finding an objective definition of mental disorder. In the first part of this essay I show that the DSM's quest for objectivity has never been successful. In the second part I argue that even if an objective definition could be found, the DSM should have no role to play in the courtroom. Today, the lawyers and forensic psychiatrists who rely on the DSM to give weight to their legal opinions and judgments are making a conceptual mistake: they conflate the concepts of disease and incapacity. Once these concepts are disentangled, it becomes apparent that the DSM (which classifies diseases) should have no business meddling with the law (which is concerned with incapacities). In the third and final part I describe the positive consequences for both parties of the divorce between the DSM and forensic psychiatry.

On January 8, 1975, the philosopher and historian Michel Foucault began his course at the Collège de France, the most prestigious and serious academic institution in France, with something unusual: he made his students laugh (Foucault 1999, pp. 3–7). He did so not by telling a joke, but by reading a psychiatric forensic case that took place in 1955. A woman had been convinced by her lover to kill her own child. In the passage that Foucault read, the psychiatrists were mostly interested in the lover. They first pointed out that this man had been a bastard child; they then described his miserable childhood and his suspicious lifestyle as an adult; they noted in particular how he liked to hang out with leftist intellectuals with revolutionary ideas, how resistant he had been to military discipline, and how he had had many mistresses, who for the most part had been low-life women. In very colorful language, these psychiatrists described in front of the jury an abject, slimy, sordid individual.

P. Singy (✉)
Union College, Schenectady, NY, USA
e-mail: pbsingy@gmail.com

P. Huneman et al. (eds.), *Classification, Disease and Evidence*, History, Philosophy and Theory of the Life Sciences 7, DOI 10.1007/978-94-017-8887-8_6, © Springer Science+Business Media Dordrecht 2015

Foucault used this forensic report and similar ones to argue that forensic psychiatry is a very peculiar type of discourse, because it has three characteristics that are not usually combined within a single discourse. Its first characteristic is power. A forensic psychiatrist has the ability to convince a jury that a defendant is mentally ill, and therefore legally irresponsible, or on the contrary mentally sane, and legally responsible. In some cases, this means the difference between life and death.

Other discourses, and above all law itself, are of course powerful as well. But they are not usually combined with the second characteristic of forensic psychiatry, which is knowledge, in the sense of scientific knowledge. Law is a rigorous and complicated formalization of a society's values; it naturally changes when those values change. "Don't Ask Don't Tell" was recently repealed in the USA, for instance, because American society changed its attitude toward homosexuality. Psychiatry, on the other hand, is supposed to be a science whose results change only because of progress in theory or the discovery of new evidence. In the same way that physics is supposed to identify objectively physical objects and their properties, psychiatry should be able to identify objectively psychiatric objects and their properties. "Objectively" is the key word here: if forensic psychiatrists claimed that they can only provide subjective, value-laden opinions about who is sick and who is sane, lawyers would not rely on them as expert witnesses. Since the early nineteenth century the legal profession has come to trust that the separation between the sane and the insane can be effectuated by psychiatrists with the tools of science.

Finally, forensic psychiatry has also, according to Foucault, a third characteristic, which a discipline like physics is lacking: forensic psychiatry can make you laugh. Why did people laugh when Foucault read them the forensic report? Without a doubt, it was because of the discrepancy between, on the one hand, the seriousness of the situation and the institutional prestige of the forensic experts, and, on the other hand, the clear lack of scientific rigor of these experts. With the help of historical hindsight, it was obvious to Foucault and his auditors that instead of providing an objective assessment of the defendant, those experts were simply translating in scientific jargon their prejudices against anything that could threaten their own conservative, middle-class, bourgeois morality.

Power, knowledge, and laughter: in the present essay I am interested in the place of the *Diagnostic and Statistical Manual of Mental Disorders* (DSM) within this Foucauldian triangle.

Diagnostic Objectivity

Since the second half of the twentieth century the DSM has drawn a *de facto* line between the normal and the pathological by listing the mental disorders officially recognized by the American Psychiatric Association, and by excluding many conditions from its pages. The consequences of this conceptual separation between health and disease cannot be overstated, not only because of the hegemony of the DSM in psychiatric practice (in the USA reimbursement by insurance companies requires a

DSM diagnosis), but also, as we will see, because the DSM has come to play a role in the courtroom. Yet, despite its clinical, forensic, and overall cultural importance, the DSM has never been able to produce a satisfactory definition of mental disorder.

In the first two DSMs (DSM-I of 1952 and DSM-II of 1968) the concept of mental disorder was not even defined. The assumption probably was that you know a mental disorder when you see one. In the 1960s and 1970s, this cavalier approach became irritating to homosexuals, who no longer accepted to be described as sick individuals. They forced psychiatrists to rethink their position on homosexuality. Eventually, a vote took place and 58 % of psychiatrists decided that homosexuality should be removed from the DSM (Bayer 1987).

The psychiatrist who was mostly responsible for the removal of homosexuality is Robert L. Spitzer. What Spitzer did in the 1970s with the DSM-III (1980) was to offer a definition of mental disorder that enabled him at the same time to exclude homosexuality and to include many conditions that were uncontroversially thought to be diseases. Although it would be interesting to parse in detail Spitzer's original definition of mental disorder, and although several small changes have been made to the definition between the DSM-III and today's DSM-5, it is clear that there has been an unchanging core of the definition, which I would state as follows: from the DSM-III until the DSM-5, a mental disorder has been defined as a *harmful dysfunction*. Something is a disorder if and only if it is at the same time a dysfunction and a cause of harm. This definition is meant to cover not just mental disorders but any medical disorder.

For instance, a heart attack is clearly a disorder because it is a dysfunction (it disturbs the function of the heart, which is to pump blood) and it causes harm. Homosexuality, on the other hand, is not a disorder because while it might be dysfunctional (if we assume that the function of sex is reproduction—more on this point later), it is not harmful. Homosexuals are often well-adjusted individuals who do not suffer from their condition *per se*. Something like teething does not meet either the requirement for something to count as a disease, but for the reverse reason than homosexuality: it is a harmful but natural, non-dysfunctional process.

With his definition of disorder as *harmful dysfunction*, Spitzer managed to remove homosexuality from the DSM without entirely challenging the validity of the rest of the disorders. The definition could in principle also serve as a conceptual test that conditions have to pass in order to be included in the DSM. This was especially important given the cultural context of the 1970s. At the time, the anti-psychiatry movement was accusing psychiatry of bringing support to the regimes around the world that were trying to oppress sexual, racial, and political minorities. Foucault is one of the most sophisticated representatives of this movement. To Spitzer's credit, he took the anti-psychiatry accusation seriously. He understood that psychiatry needed to rely on an objective definition of mental disorder that would block, or at least make difficult, the direct translation of social deviance into mental disorder.[1]

---

[1] The 1970s philosophical debate about the nature of health and disease was in great part motivated by a fear of what Christopher Boorse called the "psychiatric turn," i.e., the "strong tendency ... to debate social issues in psychiatric terms" (1975, p. 49).

And yet, the DSM definition of mental disorder has remained toothless, in part because it does not specify what "dysfunction" means. Here is a well-known example that illustrates how the lack of a specification of dysfunction can have very troublesome consequences. In 1851, Dr. Samuel A. Cartwright described a condition that he called "drapetomania," a disease that made slaves try to flee captivity (1851, pp. 707–9). Is "drapetomania" a real disease, or is it only the fantasy of a racist doctor? If we apply the DSM's current definition of mental disorder, the answer might very well be that it is a real disease, depending on how we interpret "dysfunction." Indeed, slaves who try to escape from their masters do not "function" in a racist society: they are socially dysfunctional. And they harm their masters by depriving them of their property.

The idea that a slave who tries to escape from his masters does not function in a racist society, is obviously very different from the idea that the heart does not function when it is having a heart attack. In the example of drapetomania, "dysfunction" is no longer a biological or psychological concept, but a socio-cultural one. Despite its good intentions, Spitzer's response to the anti-psychiatry critique only plays hide and seek with the thorny problem of values. The criterion of "dysfunction" gives an air of objective rigor to Spitzer's definition of mental disorder, seemingly making it less dependent upon socio-cultural values. But because "dysfunction" can in fact be interpreted in a socio-cultural manner, its objectivity is more illusory than real. Prejudices can now hide behind a veil of objectivity.

Because the current definition of mental disorder is conceptually unstable, the question "What is mental disorder?" remains very much alive today. Is racism a mental disorder or a moral flaw? Is it normal to mourn the recent loss of a spouse, or is it a symptom of depression? Should fetishism stay in the DSM or should it be removed from it? The fact that these questions and many others keep coming up indicates that we have not found a satisfactory objective criterion to separate the normal from the pathological.

The most influential people in the field do not seem overly worried by this state of affairs. This is especially clear with the paraphilias, or sexual perversions, which as I will explain below are particularly important for forensic psychiatry. Instead of trying to improve on Spitzer's well-intentioned but deficient definition, the people working on the DSM-5 have shamelessly embraced the criterion of cultural abnormality and social deviance. For them, a pervert, to put it simply, is someone who is weird. Here are two quotes from key participants in the DSM-5 that illustrate their approach to the definition of paraphilia: "The core of the paraphilia construct [is] an abnormal sexual interest. What counts as 'abnormal' is culturally relative" (Thornton 2010, p. 411); "Paraphilias are characterized by persistent, socially anomalous or deviant sexual arousal" (APA 2011). The DSM-5 itself proposes a definition of paraphilia that includes the legal criterion of consent, which obviously makes this definition directly dependent on socio-cultural values: "The term *paraphilia* denotes any intense and persistent sexual interest other than sexual interest in genital stimulation or preparatory fondling with phenotypically normal, physically mature, consenting human partners" (APA 2013, p. 685).

Since the DSM-5 is used in forensic settings, its innovations will only undermine the scientific legitimacy of forensic psychiatry even further. Forensic psychiatry is predicated upon the assumption that not all evil people are sick people. You need a forensic psychiatrist to tell you that *this* rapist suffers from a paraphilia and is not responsible, whereas *that* rapist is mentally sane and therefore responsible. If it turns out that the criterion for deciding whether someone is sick or not is that this person did something culturally abnormal, as the DSM-5 claims, then logically all criminal people will be seen as suffering from mental disorders. It is no wonder that many psychiatrists have been worried about the publication of the DSM-5.

Among these psychiatrists, there is Allen Frances, who was Spitzer's successor in charge of developing the DSM-IV. Through his blog, many articles, and books, Frances has become perhaps the fiercest and most influential critic of the future DSM-5.[2] A recurring criticism of his is that the DSM-5 will lead to false positive diagnoses, for instance by indicating that some people are paraphiliacs when really they are not.

In order to know that a diagnosis is a false positive, you logically need to know what a mental disorder is. How, then, does Frances solve the problem of the distinction between health and disease? Strangely enough, he has apparently no interest in the definition of mental disorder. As he said in an interview, "there is no definition of a mental disorder. It's bullshit. I mean, you just can't define it" (Greenberg 2010). With his dismissive attitude toward the problem of the definition of mental disorder, Frances undermines his own criticism of the DSM-5. If you do not know what a mental disorder is, if you think that matters of definition are "bullshit," then it is disingenuous to accuse the DSM-5 of leading to false positive diagnoses. Frances's debate with the DSM-5 is similar to a debate between two people who disagree about which ice cream flavor tastes better.

More problematically, by framing the debate in these subjective terms, Frances indirectly supports the DSM-5's general approach to nosology at the same time that he attacks its specific decisions. This has become directly apparent in a recent article that he co-authored with Michael First. In this article Frances explains correctly that up until the DSM-III the main criterion for something to qualify as a paraphilia was that it is "bizarre" and "unusual." These terms, "bizarre" and "unusual," were dropped starting with the DSM-III-R, because, as Spitzer explained and as Frances reports, there were concerns about their "subjectivity and unreliability" (Frances and First 2011, p. 79). Frances then explains that dropping these terms was an unfortunate decision: paraphilias *should* be defined with terms like "bizarre" and "unusual." Frances and the DSM-5 clearly disagree on what counts as bizarre, unusual, or abnormal; but they share the same fundamental assumption that the distinction between health and disease can only be made with subjective values.

Some scholars have tried to resist this trend toward subjectivity in a more lucid manner. They are continuing Spitzer's effort but push for a stricter, more objective understanding of "dysfunction" (Spitzer 1999; Wakefield and First 2003). They have usually relied on the influential work of Jerome Wakefield, who also speaks in

---

[2] See his blog at www.psychologytoday.com/blog/dsm5-in-distress

terms of harmful dysfunction but with a crucial precision regarding "dysfunction" (Wakefield 1992a, b).

According to Wakefield, a function should be understood as a mechanism selected by evolution. For instance, the heart makes a beating noise and it pumps blood. But making noise is not the function of the heart, whereas pumping blood is. Why? Because, as Wakefield explains, "it is the pumping, and not the sound, that explains why we have hearts and why hearts are structured as they are" (Wakefield 1992a, p. 236). The same holds true for psychological or behavioral characteristics. For instance, fear might have the evolutionary function of making you better prepared to face possible dangers: when you are afraid, your senses are heightened, you no longer feel pain, hunger or thirst, all your attention is concentrated on the danger and on how to avoid it. Seen from this evolutionary perspective, someone who is afraid when there is absolutely no danger has a dysfunctional type of fear and might suffer from a form of pathological phobia.

The theory of evolution is thus meant to ground the concept of "mental disorder" in biological and psychological science, and to make it less directly dependent on socio-cultural values. Now we can say, for instance, that Cartwright's "drapetomania" is not only revolting by modern cultural standards (we hope) but also scientifically wrong. From an evolutionary point of view, trying to run away from an oppressor is certainly not dysfunctional.

Wakefield's evolutionary approach is certainly attractive. Intuitively it makes sense, and politically it seems useful. But adopting Wakefield's proposal could have dangerous consequences, which follow from the fact that in practice it is extremely difficult to know what is the natural function of something.[3]

Take something as basic as sex. Here is what Spitzer wrote, in an article co-authored with Wakefield: "One does not need knowledge of evolutionary theory to recognize that the function of sexual attraction is to facilitate selection of fertile mates and behavior that leads to reproduction" (Spitzer and Wakefield 2002, p. 499). And here is what Spitzer wrote in another article: "Why do we have sexual arousal? It is obvious. Sexual arousal brings people together to have that interpersonal sex. Sexual arousal has the function of facilitating pair bonding which is facilitated by reciprocal affectionate relationships" (Spitzer 2005, p. 114). Those are of course two completely different statements about the function of sex, and this striking discrepancy illustrates how difficult it can be to identify a natural function with certainty.

And therein lies the danger. The fact that the evolutionary function of a behavior remains often indeterminate could very well encourage psychiatrists to use the theory of evolution to reinforce pre-existing prejudices.[4] John Sadler has argued for

---

[3] See for instance Derek Bolton (2008, p. 131): "for the vast majority of syndromes in the manuals we *just do not know* whether they involve failure of a natural designed function or whether they are designed or acquired strategic responses to environmental conditions, or indeed whether they are designed adaptive responses."

[4] For a general critique of evolutionary psychology and of its uses, see Kitcher (1985) and Dupré (2001).

instance that one could use Wakefield's approach to make masturbation into a disorder: "masturbation should be a disorder because it diminishes fertility in males, diminishes coital frequency, and deposits millions of potential people in various unsavory places. We are talking about reproductive compromise here" (Sadler 1999, p. 435). It is telling that Wakefield himself has had a pattern of running to the rescue of the DSM by claiming that such and such DSM category refers to an evolutionary dysfunction, even when no solid evidence could back up this claim.[5]

Instead of providing us with an objective way of distinguishing between disease and health, an approach based on the theory of evolution might only give a varnish of objectivity to what remains a subjective distinction, given that the results of evolutionary psychology are much too inconclusive to help us in concrete situations.

It seems that we have hit a wall. Since the 1960s the DSM has tried to offer an objective definition of mental disorder that would have made forensic psychiatry more scientific, less laughable; but it has failed. Today the most influential psychiatrists, like those behind the DSM-5 or their arch-enemy Allen Frances, have simply given up on that effort and have embraced subjective criteria, while more theoretically sophisticated scholars like Jerome Wakefield have provided definitions that cannot be put into practice without danger. What can be done?

Disease vs. Incapacity

So far, when facing the Foucauldian triangle, psychiatrists have assumed that the way to escape from it would be to find an objective definition of mental disorder. Psychiatrists have assumed that since lawyers are interested in knowing whether someone is sick or sane, they should find a method that could objectively determine whether someone is sick or sane, and this is what they have tried (but failed) to do through the multiple editions of the DSM.

But this assumption is based on a misunderstanding, a mixing up of the concepts of disease and incapacity. The legal system is interested in incapacities, such as not being able to think rationally or, most importantly, not being able to control oneself. The DSM is interested in diseases, such as dementia or, most importantly, the paraphilias. Obviously, there can be an overlap between an incapacity and a disease. If the overlap were 100 %, still the right question to ask in a courtroom should not be whether a defendant has a disease or not, but whether he has an

---

[5] See for instance Wakefield (1997, p. 256): "We do not have to know the details of evolution or of internal mechanisms to know ... that typical cases of thought disorder, drug dependence, mood disorders, sexual dysfunction, insomnia, anxiety disorders, learning disorders, and so on, are failures of some mechanisms to perform their designed functions; it is obvious from surface features." For a criticism of Wakefield on this point, see Demazeux (this volume; 2010); Murphy (2006, p. 44); Murphy (2011, p. 128). See also Demazeux's similar criticism of Boorse in Demazeux (2011, pp. 375–6). John Z. Sadler (1999, p. 434) rightly remarks that Wakefield's early work was more prescriptive than it is today: "It appears [Wakefield] has gone from evaluating categories for assignment of disorder status to explaining, post hoc, why the status quo is the status quo."

incapacity or not. But since, as we will see, the DSM itself admits that the overlap is far from perfect (some paraphiliacs cannot control themselves but others can, for instance), it becomes absolutely crucial in a forensic context to focus on the right issue—the issue of incapacity. This implies that whether psychiatry can find an objective definition of mental disorder or not is irrelevant for the scientific legitimacy of forensic psychiatry.

We can illustrate this problem with an example, the paraphilias. I focus on the paraphilias for a good reason: they have become forensically extremely important in the USA because of the Sexually Violent Predator laws, or SVP laws.[6] Washington was the first state to pass such a law in 1990. Since then many states have adopted similar laws, whose constitutionality has been reaffirmed twice by the US Supreme Court.

SVP laws allow for the indefinite civil commitment of people who are deemed dangerous because of their mental illness. A rapist who is deemed mentally abnormal first goes to jail, and later on continues his confinement indefinitely in a psychiatric institution until he is "cured." The assumption that justifies the difference of treatment between this "mentally insane" rapist and an ordinary "clinically sane" rapist is that only the former is a "predator," i.e., some kind of animal whose very nature pushes him to rape. To release him would be akin to releasing a tiger in a city. A "clinically sane" rapist, on the other hand, is an evil person who can control himself if he wants to. The Supreme Court has made clear that SVP laws would be unconstitutional if it were not for the existence of a disease in the defendant.

Although the presence of a mental disease is crucial for the application of SVP laws, this does not mean that the DSM itself is the ultimate arbiter in SVP decisions. In the Supreme Court case of *Kansas v. Hendricks* (1997), the majority clearly stated that the legal and the psychiatric concepts of mental disorder are not and do not need to be equivalent: "The legal definitions of 'insanity' and 'competency'… vary substantially from their psychiatric counterparts…. Legal definitions … need not mirror those advanced by the medical profession."

Nevertheless, as a matter of fact the DSM clearly matters in the courtroom. A defendant who receives a diagnosis that has been vetted by the American Psychiatric Association is more likely to be seen as "really" mentally ill. Always in *Kansas v. Hendricks*, the decision to commit the pedophile Hendricks is upheld by the Supreme Court on the ground that "the mental health professionals who evaluated Hendricks diagnosed him as suffering from pedophilia, *a condition the psychiatric profession itself classifies as a serious mental disorder*."[7] The DSM, in other words, can give legitimacy to what otherwise could be seen as a dubious diagnosis. In the eyes of a jury, the sexual criminal who receives an official DSM diagnosis is probably more likely to be a sexual predator than the sexual criminal who receives a diagnosis that has not been vetted by the American Psychiatric Association.

What is it about mentally ill people, and only them, that can make them qualify for civil commitment? People with cancer, people without a college degree, people

---

[6] On SVP laws, see in particular Janus (2009).

[7] This particular point is reaffirmed in *Kansas v. Crane* (2002).

who have a history of crimes, or people with a mustache, for instance, cannot possibly qualify for SVP commitment—only mentally ill people can. Why is that?

In the case of the paraphilias, a psychiatric diagnosis is legally relevant only inasmuch as it indicates either the presence or the lack of self-control. *Kansas v. Hendricks* is very clear on this point: "The precommitment requirement of a 'mental abnormality' or 'personality disorder' ... narrows the class of persons eligible for confinement to those who are unable to control their dangerousness."[8] SVP Lawyers are interested in mental disorders only insofar as these disorders signal a lack of self-control. To have a psychiatric diagnosis is a necessary but not sufficient condition for being civilly committed. Otherwise, any depressed criminal would qualify for civil confinement, and this is clearly not the case. The legal system is interested in incapacities, not diseases *per se*.

Contrast this with what the DSM itself says regarding the capacity of self-control: "a diagnosis does not carry any necessary implications regarding ... the individual's degree of control over behaviors that may be associated with the disorder. Even when diminished control over one's behavior is a feature of the disorder, having the diagnosis in itself does not demonstrate that a particular individual is (or was) unable to control his or her behavior at a particular time" (APA 2013, p. 25). As Bruce Winick astutely remarked, the DSM in fact never claims that someone with a mental disorder is unable to control himself, including when you would most expect it: "The language in DSM-IV describing even the impulse control disorders and sexual disorders—conditions involving repetitive criminal and sometimes violent behavior—suggests a failure on the part of the individual to resist strong impulses or urges, rather than an inability to do so" (Winick 1995, p. 579 (n. 189)).

Since the DSM remains agnostic about lack of self-control, and since it is precisely this incapacity that is at stake in the courtroom in cases of paraphilias, having a DSM diagnosis should be forensically irrelevant. One would think that this is what is implied in the previously quoted passage from the DSM-5. Yet this passage is rather oddly preceded by another one, which contradicts it directly: "When used appropriately, diagnoses and diagnostic information can assist legal decision makers in their determinations. For example, when the presence of a mental disorder is the predicate for a subsequent legal determination (e.g., involuntary civil commitment), the use of an established system of diagnosis enhances the value and reliability of the determination" (APA 2013, p. 25). The DSM claims on the same page that a diagnosis does not imply lack of self-control, and that a diagnosis helps decide an issue that is fundamentally about self-control.

The same type of contradiction appears in an official document published in 1992 by Seymour Halleck and his collaborators, entitled "The Use of Psychiatric Diagnoses in the Legal Process: Task Force Report of the American Psychiatric Association." The authors of this report tell us, very correctly, that a diagnosis "does not inform the legal decision maker about the actual impairment of a particular patient," and when they turn to the specific but crucial question of self-control, they admit frankly that "no science of volition exists." However, despite these cautionary

---

[8] For a criticism of this statement, see Singy (forthcoming).

statements, the authors claim that "psychiatrists' ability to make prognostic or retrospective judgments flow from diagnoses. Correctly diagnosing the patient is an essential step in any such evaluation" (Halleck et al. 1992, pp. 491, 494, 489).

We can sum up the convoluted relation between the law and the DSM in the following way: the law says that in order to be civilly committed one needs to have a mental disorder, and, logically enough, forensic psychiatrists use the DSM to tell if someone has a mental disorder; but the law is interested in knowing whether someone suffers from a mental disorder only because it is interested in whether someone has an incapacity, and the DSM warns that having a mental disorder does not imply the existence of an incapacity.

The DSM needs to modify its section on the use of the DSM in forensic settings. Since for the law what matters is incapacity, and since the DSM cannot decide anything on this question, any use of a DSM diagnosis in the courtroom is scientifically illegitimate. There are irreconcilable differences between the DSM and forensic psychiatry, and it is time for them to get a divorce.[9] The DSM needs to be a clinical tool, and nothing else.

Forensic Psychiatry Without the DSM, and Vice Versa

The effects of the divorce between the DSM and forensic psychiatry would be beneficial for both parties.

For forensic psychiatry, psychiatrists would be forced to focus on what matters legally: not the presence or absence of disease, but the presence or absence of an incapacity. Obviously, this might make the task of forensic psychiatry much more difficult than it is now. The determination of the presence or absence of self-control (arguably the most important capacity in a forensic context) is for instance an extremely difficult task. We could probably go one step further and argue that it necessarily escapes scientific research, since science is deterministic, and the issue of self-control is ultimately an issue about free will. Raymond Saleilles, the famous constitutional lawyer of the late nineteenth century, put it simply: "Science and observation discover only causes and effects. But free will consists in making a breach into the causality principle" (Saleilles 1898, p. 80). Science cannot prove that there is free will, since this would contradict the deterministic methodological principle upon which most of science is founded. And neither can it prove that there is no free will, since determinism is not something that science discovers, but a metaphysical assumption that makes it possible. As legal scholar Michael S. Moore rightly remarked, "What psychiatry essentially lacked—and still lacks—was any reconciliation of its own deterministic assumptions with the concept of responsibility" (Moore 1980, p. 41). This is a problem that has been noted by the APA itself in its

---

[9] This would mark the end of a relationship that in fact has been going south for a while. Since the DSM-III the successive editions of the DSM have increasingly warned against the problems associated with its forensic use. See Shuman (2002, pp. 217–8).

1983 official statement on the insanity defense: "The line between an irresistible impulse and an impulse not resisted is probably no sharper than that between twilight and dusk. Psychiatry is a deterministic discipline that views all human behavior as, to a large extent, 'caused'. The concept of volition is the subject of some disagreement among psychiatrists" (APA 1983, p. 685).

To bypass the epistemological difficulties marring the scientific assessment of free will, forensic psychiatrists often replace the volitional test with an intellectual test. Instead of trying to know whether a person acted freely or was determined by his disease, they evaluate whether the person is able to appreciate the wrongfulness of his conduct. What really matters legally remains the question of will, of self-control, but the assumption, stated explicitly by the APA, is that "there is considerable overlap between a psychotic person's defective understanding or appreciation and his ability to control his behavior. Most psychotic persons who fail a volitional test for insanity will also fail a cognitive-type test when such a test is applied to their behavior, thus rendering the volitional test superfluous in judging them" (APA 1983, p. 685).

This is a rather odd statement, given that it follows the APA's previously quoted concession that psychiatry is a deterministic discipline and cannot decide whether someone suffers from volitional impairment. In order to claim that the results of a cognitive test overlap with the results of a volitional test, one obviously needs a trustworthy volitional test. How can one claim that A overlaps with B if one does not know what B is?

Banning the DSM from the courtroom would certainly make the task of forensic psychiatrists more difficult, and perhaps even impossible. At the very least, it should help make visible the scientific weakness of expert reports written by forensic psychiatrists. I, for one, think this would be an excellent thing. Forensic psychiatry can serve the humane and laudable function of helping the helpless, and it does not necessarily have the goal of reinforcing the punitive effect of the law. But with great power comes great responsibility, and one of the first responsibilities of forensic psychiatry should be to be honest about the epistemological difficulties that it faces.

The divorce between the DSM and forensic psychiatry would also be consequential for the DSM itself. It would make necessary to have a second look at all the DSM diagnostic categories and ask if they still have a *raison d'être* outside the forensic context.

For all the diagnoses that have never been particularly relevant forensically (such as depression), the divorce between the DSM and forensic psychiatry would not have any direct effect. For the other categories, however, some major changes to the DSM could be necessary. This is the case with the paraphilias, for instance, which are essentially forensic concepts, as their history clearly reveals.

The old name for paraphilia was "sexual perversion." All the terms that refer to sexual perversions, including the expression "sexual perversion" itself, have been invented in the second half of the nineteenth century, usually by German forensic psychiatrists. The emergence of this new lexical field is the sign of an entirely new way of thinking about sex, which we call "sexuality" (also a nineteenth-century word). Most importantly, these new concepts appeared in a forensic context:

psychiatrists began to talk about sadism, homosexuality, exhibitionism, fetishism, etc., not because they were concerned with helping patients in distress, but because they wanted to determine whether defendants were responsible for crimes.

For instance, the first important case of perversion dates from 1849 and was about a French soldier who had sex with corpses and tore them to pieces: the sergeant François Bertrand. The question that was asked during his trial was: did this man voluntarily desecrate corpses, or was he driven by an irresistible instinct? Psychiatrists thought that since he was young, good-looking and intelligent, if he had wanted he could have had sex with living women. The only possible explanation for his crime was therefore, according to these psychiatrists, that he suffered from a perversion of the sexual instinct.

A few years later the same type of question was raised about people who had sex with people of the same sex. This time it took place in Germany rather than in France, because sodomy was illegal in Germany but not in France. German psychiatrists determined that some people simply cannot help desiring people of the same sex. It is not a choice: it is part of their nature, they are a different kind of people than normal people. These German psychiatrists argued that "homosexuals," as they came to be known, should not be deemed legally responsible for the crime of sodomy.

All the paraphilias that have ended up in the DSM have a similar forensic origin. Historically, it has always been the same pattern: first, a crime is committed (rape, sodomy, exhibitionism, etc.), and then the issue of responsibility is raised: did the defendant commit this crime voluntarily, or could he not have helped himself because it was part of his nature? Today it is still within a forensic context that new paraphilias are created. Take for instance the fierce debate about "hebephilia," a paraphilia characterized by sexual attraction to pubescent children. "Hebephilia" was suggested for inclusion in the DSM-5 (Blanchard et al. 2009). Although the validity and usefulness of this category has been repeatedly criticized (Franklin 2010; Wakefield 2011; Singy forthcoming), and although the category was ultimately rejected by the APA's Board of Trustees, "hebephilia" is nevertheless regularly invoked in the courtroom. Today, like in the nineteenth century, the paraphilias are born in the courtroom, not on the couch.

Since, as I have argued above, I think the DSM should break off its relation with forensic psychiatry, the forensic role of the paraphilias would be a good reason to suspect that they should be removed from the DSM. Allen Frances and Michael First, two of the main architects behind the DSM-IV, have argued something similar about hebephilia: hebephilia "arose, not out of psychiatry, but rather to meet a perceived need in the correctional system. This solution represents a misuse of the diagnostic system and of psychiatry" (Frances and First 2011, p. 84).

The same exact reasoning should apply to all the current paraphilias. I suspect that if Frances and First do not reach this conclusion, it is because they are not familiar with the history of paraphilias. Frances was recently challenged by Andrew Hinderliter to explain why he did not try to remove the paraphilias from the DSM when he was the chair of the DSM-IV task force. On his blog Frances acknowledged that the paraphilias are a problematic section in the DSM, but he explained that he decided to keep them because of "the unknowable risks and inconveniences of … a radical break with longstanding diagnostic traditions" (Frances, February 2, 2011a). Since Hinderliter was understandably not satisfied with this explanation, Frances explained further that "When we say a diagnosis has been included in DSM because of 'historical tradition', this means it has been included because it has been a focus of clinical attention in the past. The DSM is first and foremost a tool for clinicians to assist them in their clinical practice" (Frances, February 3, 2011b).

Frances got his history wrong: the paraphilias were not included in the DSM because they have been a focus of clinical attention in the past, but because they have been a focus of forensic attention. Frances's reasoning for not including hebephilia in the DSM-5 seems perfectly sound to me, but it should be applied to the current paraphilias as well. All of them should be removed, because all of them are forensic concepts, and the DSM should not play any forensic role.

But wouldn't there be at least a clinical reason for keeping the paraphilias in the DSM? The paraphilias have a forensic *origin*, but that does not mean that they need to function nowadays as forensic concepts. Over time, they might have been transformed into clinical concepts. What Sadler has called the "victimizing paraphilias," like pedophilia or sadism, clearly remain forensic concepts and should therefore be excluded from the DSM if my previous argument is correct (Sadler 2005, p. 210). But what about something like fetishism, for instance? If a fetishist is miserable, shouldn't psychiatrists try to cure him of his fetishism? Can't a diagnosis of paraphilia make clinical sense in some cases?

Here we need to ask ourselves: what exactly does this fetishist suffer from? Compare with an anorexic: the anorexic clearly suffers from her condition, in fact she might die from it. If paraphiliacs suffer, on the other hand, it is not because of their sexual preference *per se*. It is because of how people react to their sexual preference. For instance, it is of course very possible that a homosexual would suffer from depression, and that it would not be the case if he were heterosexual. But depression is not caused by homosexuality itself; it is caused by the discrimination that a homosexual might feel in a homophobic society. Homosexuality is not the problem: the problem is society's reaction to homosexuality.

The situation is analogous to African-Americans who are depressed as a result of living in a racist environment, or to women who are depressed as a result of living in a misogynistic environment (see Moser and Kleinplatz 2005, p. 101). Psychiatrists might want to help African-Americans and women cope with their depression, but they will not try to cure them of their race or womanhood, if this were possible. The same should hold true for non-victimizing paraphiliacs, and this is why the inclusion of the paraphilias in the DSM does not make more sense clinically than it does forensically.

Conclusion

Facing the embarrassment symbolized by the Foucauldian triangle of power/ knowledge/laughter, psychiatrists have tried to reinforce the objectivity of the "knowledge" side of the triangle. This, they assumed, would establish the scientific seriousness of their discipline and protect them from the laughter of their critics. We can certainly applaud this effort, even if the recent publication of the DSM-5 makes clear that it did not yield any solid result. More than ever, psychiatry is an epistemologically dubious science.

What I have suggested in this essay is a different strategy to escape the Foucauldian triangle. The laughter of Foucault's audience originated in the gap between the great power enjoyed by psychiatrists and the low scientificity of their claims. Since raising the scientificity of psychiatry has failed, one must lower its power. I have argued in particular that the DSM should no longer meddle with the law: it must limit the exercise of its power to the clinical setting. This strategy might seem a bit spineless, intellectually speaking: instead of raising to the challenge posed by the laughter of critics and try to make psychiatric nosology into an epistemologically sound science, one attempts to defuse this laughter by avoiding situations where psychiatry's lack of scientific rigor can have deadly consequences. There is however a good conceptual reason to adopt this strategy: as I have tried to show, the DSM and the law are respectively concerned with overlapping but distinct concepts. The DSM itself already distinguishes, though not always coherently, between diseases (relevant to the DSM) and incapacities (relevant to the law). It only fails to reach the proper conclusion and refuses to let go of the power it still unduly enjoys in the courtroom.

References

American Psychiatric Association (1983) American Psychiatric Association statement on the insanity defense. Am J Psychiatry 140:681–688

American Psychiatric Association (2011) Hypersexual disorder, rationale. From http://www.dsm5. org/proposedrevision/Pages/proposedrevision.aspx?rid=415# (Downloaded on October 6, 2011)

American Psychiatric Association (2013) Diagnostic and statistical manual of mental disorders, 5th edn. Author, Arlington

Bayer R (1987) Homosexuality and American psychiatry. Princeton University Press, Princeton

Blanchard R, Lykins AD, Wherrett D, Kuban ME, Cantor JM, Blak T, Dickey R, Klassen PE (2009) Pedophilia, hebephilia, and the *DSM-V*. Arch Sex Behav 38:335–350

Bolton D (2008) What is mental disorder? An essay in philosophy, science, and values. Oxford University Press, Oxford

Boorse C (1975) On the distinction between disease and illness. Philos Public Aff 5:49–68

Cartwright SA (1851) Report on the diseases and physical peculiarities of the Negro race. New-Orleans Med Surg J 7:691–715

Demazeux S (2010) Le concept de fonction dans le discours psychiatrique contemporain. Matière Prem 1:29–72

Demazeux S (2011) Le lit de Procuste du DSM-III. Classification psychiatrique, standardisation clinique et ontologie médicale. Université Paris 1 – Panthéon-Sorbonne, Paris

Dupré J (2001) Human nature and the limits of science. Clarendon, Oxford

Foucault M (1999) Les Anormaux. Seuil/Gallimard, Paris

Frances A (2011a) A conservative approach to diagnosis grandfathers in weak links. Blog DSM5 in Distress. www.psychologytoday.com/blog/dsm5-in-distress, 2 Feb 2011

Frances A (2011b) Sunsetting psychiatric diagnoses. Blog DSM5 in Distress. www.psychologyto-day.com/blog/dsm5-in-distress, 3 Feb 2011

Frances A, First MB (2011) Hebephilia is not a mental disorder in DSM-IV-TR and should not become one in DSM-5. J Am Acad Psychiatry Law 39:78–85

Franklin K (2010) Hebephilia: quintessence of diagnostic pretextuality. Behav Sci Law 28:751–768

Greenberg G (2010) Inside the battle to define mental illness. Wired, 27 Dec 2010

Halleck SL, Hoge SK, Miller RD, Sadoff RL, Halleck NH (1992) The use of psychiatric diagnoses in the legal process: task force report of the American Psychiatric Association. Bull Am Acad Psychiatry Law 20:481–499

Janus ES (2009) Failure to protect: America's sexual predator laws and the rise of the preventive state. Cornell University Press, Ithaca

Kitcher P (1985) Vaulting ambition: sociobiology and the quest for human nature. The MIT Press, Cambridge, MA

Moore MS (1980) Legal conceptions of mental illness. In: Brody BA, Engelhardt HT (eds) Mental illness: law and public policy. D. Reidel Publishing Company, Dordrecht

Moser C, Kleinplatz PJ (2005) *DSM-IV-TR* and the paraphilias: an argument for removal. J Psychol Hum Sex 17:91–109

Murphy D (2006) Psychiatry in the scientific image. The MIT Press, Cambridge, MA

Murphy D (2011) Thinking about the foundations of psychiatry: an interview with philosopher Dominic Murphy. Psychiatrie Sci Hum Neurosci 9:125–30

Sadler JZ (1999) Horsefeathers: a commentary on 'evolutionary versus prototype analyses of the concept of disorder'. J Abnorm Psychol 108:433–437

Sadler JZ (2005) Values and psychiatric diagnosis. Oxford University Press, Oxford

Saleilles R (1898) L'individualisation de la peine. Etude de criminalité sociale. Félix Alcan, Paris

Shuman DW (2002) Softened science in the courtroom: forensic implications of a value-laden classification. In: Sadler JZ (ed) Descriptions and prescriptions: values, mental disorders, and the DSMs. The Johns Hopkins University Press, Baltimore

Singy P (forthcoming) Danger and difference: the stakes of hebephilia. In: The DSM-5 in perspective: philosophical reflections on the psychiatric babel. Springer

Spitzer RL (1999) Harmful dysfunction and the DSM definition of mental disorder. J Abnorm Psychol 108(3):430–32

Spitzer RL (2005) Sexual and gender identity disorders: discussion of questions for *DSM-V*. J Psychol Hum Sex 17:111–116

Spitzer RL, Wakefield JC (2002) Why pedophilia is a disorder of sexual attraction—at least sometimes. Arch Sex Behav 31:499–500

Supreme Court. Kansas v. Hendricks, 521 U.S. 346 (1997)

Supreme Court. Kansas v. Crane, 534 U.S. 407 (2002)

Thornton D (2010) Evidence regarding the need for a diagnostic category for a coercive paraphilia. Arch Sex Behav 39:411–418

Wakefield JC (1992a) Disorder as harmful dysfunction: a conceptual critique of DSM-III-R's definition of mental disorder. Psychol Rev 99:232–247

Wakefield JC (1992b) The concept of mental disorder: on the boundary between biological facts and social values. Am Psychol 47:373–388

Wakefield JC (1997) Normal inability versus pathological disability: why Ossorio's definition of mental disorder is not sufficient. Clin Psychol 4:249–258

Wakefield JC (2011) DSM-5 proposed diagnostic criteria for sexual paraphilias: tensions between diagnostic validity and forensic utility. Int J Law Psychiatry 34:195–209

Wakefield JC, First MB (2003) Clarifying the distinction between disorder and nondisorder. In: Phillips KA, First MB, Pincus HA (eds) Advancing DSM dilemmas in psychiatric diagnosis. American Psychiatric Association, Arlington

Winick BJ (1995) Ambiguities in the legal meaning and significance of mental illness. Psychol Public Policy Law 1:534–611

Defining Genetic Disease

Catherine Dekeuwer

Abstract The concept of genetic disease refers to the idea that one or more genes are the cause of disease. Under this definition, problems arise when it comes to the use of the term "cause". Moreover, genes alone cannot explain the development of a disease; environmental causes are also at play. So when is giving primary importance to genetic causation justified?

In his paper "The Concept of Genetic Disease" (2004), David Magnus differentiates three competing concepts of genetic disease related to three approaches of causality. He concludes that none is really acceptable. Finding insufficiencies in each approach, he ultimately adopts a definition of genetic disease that arises from medical uses. In this contribution, I will specify the three conceptions of genetic causality and defend the idea that they function *together* in the definition of a disease as a genetic disease.

I wish to contribute to the philosophical reflection on the definition of genetic disease (Hull 1979; Hesslow 1984; Sterelny and Kitcher 1988; Gifford 1990; Gannett 1999; Sober 2001; Kitcher 2003; Magnus 2004) by showing that three criteria are necessary, together, in order to consider a disease to be genetic. Before entering into the subject, however, I would like to point out two things. First, the concept of disease itself is difficult to define. Does this difficulty prevent, however, any attempt at defining certain diseases as genetic? On the contrary, I think that the analyses presented here could contribute to the general discussion of what disease is. Second, there is the problem of figuring out where to begin when it comes to

C. Dekeuwer, Ph.D. (✉)
Faculté de philosophie, Université Lyon 3, Lyon, France

Institut de Recherches Philosophiques de Lyon, Université Lyon 3,
Lyon, France
e-mail: catherine.dekeuwer@univ-lyon3.fr

P. Huneman et al. (eds.), *Classification, Disease and Evidence*, History,
Philosophy and Theory of the Life Sciences 7, DOI 10.1007/978-94-017-8887-8_7,
© Springer Science+Business Media Dordrecht 2015

defining genetic diseases. The concept of genetic disease is based on the idea that one or several genes cause it; what is at issue is how we conceive these "causes". Furthermore, the development of a pathology cannot be attributed to genes alone. The problem then is to know why, for some diseases, researchers assign greater importance to genetic causes than to environmental ones. Does finding a solution to this problem require a critical re-evaluation of the ways the term "genetic disease" is currently used? Or does it require instead a conceptual analysis of genetic causality that would clearly delineate the line between genetic and non-genetic diseases? And how does the history of medicine and biology factor into this research? It seems to me that the uses of the expression "genetic disease" evoke both a series of problems that doctors seek to resolve, as well as the concepts they use to achieve such solutions. Understanding the notion of genetic disease therefore demands a twofold understanding of the term's concept as well as its history.

One of David Magnus's articles entitled "The Concept of Genetic Disease" (Magnus 2004) serves as the basic frame of reference for my presentation of the three approaches to genetic causality. Magnus distinguishes three competing concepts of genetic disease, which correspond to three approaches to causality. According to Magnus, none of them is truly satisfactory, which leads him to consider the uses of the concept in order to understand why genetic causality is often privileged in studies of certain diseases. However, whereas Magnus critiques the insufficiencies of each approach and reduces the meaning of genetic disease to its medical applications, I am claiming here that the idea of these three conceptions of causality, when rendered more precisely, function together within the definition of certain diseases as genetic disease.

Problems Involving Genetic Causality

Magnus distinguishes three approaches to genetic causality, each with its own set of issues: it is thus impossible, according to Magnus, to settle on one definition of genetic disease in the absence of any firm epistemological foundation. In the first part of this chapter, I would like to review these three approaches and discuss their critiques.

First, a disease is genetic if it results from the direct causal action of one or more genes. In order to explain the concept of direct causal action, Magnus uses an article by Fred Gifford (1990, p. 329) in which he argues that in order to be considered genetic, "the trait must be the specific effect of some genetic cause, that the trait must be described or individuated in such a way that it is properly matched to what the gene causes specifically". Magnus does not even take the time to refine his critique of this concept of causality: for him, the development of a disease is so complex that it is useless to try to find its direct genetic causes. However, on this point he confuses direct causality with specific causality. Gifford does not mention direct causality; he uses the concept of specific effect defined as a correspondence established by researchers between, on one hand, what the gene specifically causes, and on the other, the phenotype as it is described and individuated. The effect is specific if the modification of a gene has some effect on the considered trait but not on other traits. The phenotype must be individuated (i.e. correspond to one precise unit of description), neither too broad nor too narrow. Gifford first takes the example of language acquisition. If genes do have an effect on aptitude for learning a language, the ability to learn French is not genetic: the trait is too specifically individuated. Another example: hypercholesterolemia cannot be considered a genetic disease. It is necessary to distinguish familial and sporadic forms of this disease, by individuating the reported phenotype from a genetic cause. This perspective has the advantage of directing our attention to the need to clearly identify and delineate phenotypes for which genetic causes may be sought. The absence of a clear description and individuation for schizophrenia, for instance, has made research on its genetic determinants quite difficult (Maziade et al. 2003).

The concept of specific effect raises questions related to molecular pathology. Gifford explains that it would be better to give less importance to the question of knowing if a trait is genetic in order to focus instead on "what are the steps involved in the biosynthetic or developmental pathway?" (Gifford 1990, p. 328). And yet, "Molecular pathology seeks to explain why a given genetic change should result in a particular clinical phenotype (f). Molecular pathology requires us to work out the effect of a mutation on the quantity or function of the gene product, and to explain why the change is or is not pathogenic for any particular cell, tissue or stage of development" (Strachan and Read 2004, p. 418). Molecular pathology aims to answer this type of question: "Why should loss function of the FMR1 protein, involved in transporting RNA from nucleus to cytoplasm, cause mental retardation and macro-orchidism (Fragile-X syndrome)?" (Ibid., p. 416.) In this perspective, a gene would be a cause of a disease in the sense that the identification of a DNA sequence would allow different steps in the process that leads to an individual becoming ill to be specifically explained. This approach to causality (once it is properly understood) relies then on genes' specific influence on the ways in which organisms develop. It falls under the category of an explicative approach as it relates to the biological individual where a causal history of specific effects produced by one or several genes is retraced.

The second relevant approach to causality when it comes to the definition of genetic disease relates to populations rather than individuals. A disease is genetic if "in that population, the covariance of the trait with some genetic factor(s) is greater than the covariance of the trait with other (non-genetic) factors" (Magnus 2004, p. 235). Magnus again uses one of Gifford's criteria: "trait is genetic (with respect to population P) if it is genetic factors which 'make the difference' between those individuals with the trait and the rest of population P" (Gifford 1990, p. 333). This concept of cause is statistical and only has meaning relative to populations. Analysis of variance is effectively used to measure the causal contribution of genetic and environmental factors in a given population. According to Magnus, this approach is not convincing either, precisely because it is population-relative: as a function of the studied population, a factor can be considered genetic or not. Identification of a genetic factor of a disease thus depends on the population selected for the study, which makes the concept of genetic disease too relative. Magnus makes a close study of one proposed by Hesslow (1984) and takes the example of a village well contaminated by a pathogen. Only half the villagers fall ill and the researchers thus assume that their genes confer resistance to the pathogen. On the scale of the village, the covariance of genetic factors of the disease is close to 1, whereas the covariance of the disease and the pathogen is close to 0.5. The disease is, relative to the village's population, genetic. On the planetary scale however, the disease would probably be environmental. For Magnus, this type of statistical approach to causality is not a good one for defining certain diseases as genetic. I would like to add here that evidence of a risk factor is precisely not the identification of a cause.

It is helpful to delve deeper into this critique and distinguish several statistical approaches to genetic causes. Magnus refers to a disease to which the villagers would be more or less susceptible. These are association studies that allow for the identification of the genetic terrain's determinants, but not those of genetic diseases. The notion of the terrain is relevant when the goal is to find out the genetic determinants that explain certain individuals' resistance to diseases with an environmental etiology, such as a virus. Association studies are thus adaptive: they aim to correlate a difference in allelic frequency for the same locus among non-related subjects who are or are not affected by the pathology. An allele is "associated" with a disease if it is present more frequently among the ill than among the unaffected. It confers an increased risk of disease compared with the general population, expressed by the relative risk. The choice of reference populations is thus quite crucial, but this is not, in my mind, the main problem with these association studies. Finding out why certain individuals are, for example, more susceptible to viral diseases or cancers than others are does indicate some susceptibility factors, but it would be quite difficult to justify the claim that these diseases are genetic. It would be difficult to defend the claim that AIDS is a genetic disease even if there are reported cases of resistance to HIV and even if we could identify genetic factors associated with resistance to HIV. This is where the confusion lies in the field of genetics and genetic disease.

Association studies are especially relevant for the study of multifactorial diseases with complex heredity such as asthma, diabetes, or certain forms of Azheimer's.[1] For these diseases, a familial aggregation is often observed, but the mode of transmission often remains unknown. Their study requires a method for researching these diseases' genetic determinants without knowing their mode of transmission, and the association study responds to this requirement. In 1993, such association studies demonstrated a strong statistical correlation between a polymorphism of the APOE gene (apolipoprotein E), named e4, and the most frequent forms of Alzheimer's disease, which are non-Mendelian forms. This correlation between the e4 allele of the APOE gene and the sporadic form of Alzheimer's disease has since been confirmed by independent studies; it at least

[1] These multifactorial diseases are often also multigenic diseases: many genes work together in the development of the pathology. Campion (2001), Strachan and Read (2004), Feingold (2005).

holds true for populations that descend from common European ancestors. It has been calculated that Alzheimer's disease is two to three times more frequent in the group of heterozygote individuals that carry an e4 allele than it is in the general population. The disease is 9–15 times more frequent for homozygotes, which carry this allele in two copies. The APOE gene is thus referred to as a gene of susceptibility for sporadic forms of Alzheimer's: more than half of people with the disease carry the e4 allele, but many who have this allele will never develop the disease. Moreover, the absence of this allele in a person does not mean that he or she will not develop Alzheimer's. Yet these associations are often weak, the studies difficult to reproduce, and researchers struggle to define the physiopathological meaning of an association between a marker allele and an increased disease frequency. The risk factor often lacks any biological meaning, which makes it difficult or even impossible to interpret it as a genetic cause. For example, researchers today are trying to associate markers called SNP (Single Nucleotide Polymorphism) with complex pathologies, but they know already that these markers have no biological function. If the genetic markers do not intervene in the appearance of pathologies, it is difficult to place multifactorial diseases in the genetic disease category.

Genetic linkage studies involve diseases of Mendelian heredity for which alterations of one gene have a major and specific effect on the phenotype (Huntington's disease, cystic fibrosis and hereditary breast and ovarian cancer are all examples). These studies rely on the co-segregation of a phenotype and certain alleles in families. For a monogenic disease like Huntington's, for example, some genetic markers have been identified via linkage studies. They have made it possible to predict which individual in a family is at risk; genetic tests were first performed in 1986. At that time, however, the protein coded by the gene near the marker and the molecular mechanisms of the disease were not identified. Linkage analysis does not therefore identify the "gene that causes the disease"; only the marker is identified. In 1993 a gene was identified and research on the Huntington gene began.

To summarize: linkage analyses localize genes involved in the appearance of monogenic diseases whose mode transmission is known. Statistical analysis, in this case, relies on knowing the disease's mode of transmission, which gives meaning to the notion of "genetic" disease. However, as in the case of association studies, as long as the gene's function is unknown, it is hard to extrapolate a causal relationship from a link. All of these examples serve to show that the statistical approach alone does not give us a way to understand pathological processes and only sometimes allows us to identify risk factors. It does lead though to tools which allow us to identify people who are predisposed to certain diseases. And finally, it underscores the importance of the hereditary aspect of diseases considered genetic.

According to Magnus, the example of the contamination of a well shows that we intuitively consider the pathogenic agent to be the disease's cause because we believe we are capable of acting on this cause (by cleaning the well water for instance). This third approach to causality is no longer a matter of scientific arguments but rather of interests involving medical treatment that explains why certain diseases are considered genetic. From this perspective, a disease is genetic when its genetic determinants appear easier to manipulate than its environmental ones in order to treat or prevent the disease. Magnus is careful to state, however, that there is no such thing as genetic therapy; the therapeutic justification of this instrumental approach is therefore largely compromised. Moreover, certain diseases that are said to be genetic are medically treated with environmental modifications. A special diet, for example, is enough to prevent the appearance of symptoms characteristic of phenylketonuria. Yet should this instrumental approach be reduced to its preventive dimension? Diseases would then be considered as genetic when a genetic test could be put to market that allows people to predict who is and is not at risk of the disease. This solution clearly shows the current limits of medical care for the great majority of diseases we refer to as genetic: most often, a genetic test prevents the disease by allowing for the selection of births.

Magnus considers each of these concepts of causality in turn before rejecting them all. But why not consider these three approaches to causality as all contributing together to the definition of a disease as genetic? These concepts do not seem to be competing: they refer to three aspects of medicine. In medical practice, genetic research effectively leads to specific care regimes for individuals and families. Research toward a better understanding of diseases involves localizing genetic determinants as well as explaining their physio-pathological role.

In order to better understand why it is so difficult to define one concept of genetic disease, an important distinction must be made. Lenny Moss (2004) distinguishes two concepts of the gene: the gene-P (for prediction) and the gene-D (for development). The first concept appears in the context of linkage studies that identify markers or "disease genes". In this case, the link between the gene and the disease is strong enough to make predictions for it, even if nothing is known about the gene's specific mode of action. This concept thus has an essentially instrumental value: the gene is defined by its relationship to a phenotype, but knowing the DNA sequence or the way the gene behaves is not necessarily to make a prediction as to who will be ill in a family. On the other hand, prediction allows for prevention. The D-gene is defined as a resource for development. This resource is in itself indeterminate relative to the phenotype. In this last case, the DNA sequence is known and simply considered an element of developmental processes without reference to a particular phenotype. The same gene can be considered as either a P-gene or a D-gene. For example, certain versions of the BRCA1 gene are correlated to a heightened risk of breast and ovarian cancers. In this case, BRCA1 is considered to be a P-gene. But when the BRCA1 gene is not considered as the "breast cancer gene", but rather as a model for the protein synthesis present in several cells and tissues, then it is a D-gene. These proteins can be studied in each cell and tissue context without taking the phenotype (breast cancer) into account. If the P-gene is a prediction tool used in order to obtain a medical or economic benefit, then the D-gene is of explanatory value.

With these conceptual distinctions, it is now important to develop some examples that lay out the argument for a common functioning of these three approaches to causality in order to come up with a definition of genetic disease.

The Concept of Molecular Disease

The concept of genetic disease must be understood within the framework of research on molecular pathology. Here I would like to explain this concept in more detail in addition to pointing out how it is related to medical research on the hereditary transmission of diseases. Magnus cites Linus Pauling's work and recalls the importance of the concept of "molecular disease" that appeared in his famous 1949 article (Pauling et al. 1949, cf. Feldman and Tauber 1997). But Magnus does not pay enough attention to the specificity of Pauling's approach and its connection to the geneticist James Neel's work (1949).

To better understand the concept of genetic disease at play here, it is useful to return to a simple question. Why is it that, although certain individuals only suffer from sickle-cell disease at high altitudes, this disease is classified as genetic rather than environmental? The answer lies partly in a molecular analysis of sickle-cell disease carried out in the late 1940s and 1950s' (Pauling et al. 1949; Ingram 1957; Ingram and Stretton 1959). At the beginning of the 1949 article, Pauling's team points out that sickle-cell is characterized by severe anemia, which results from cellular abnormalities. The anemic trait is a less-severe form of the disease, which is not felt by most individuals who have it. This disease is defined, at the cellular level, by two characteristics: the special "sickle" form which red blood cells take, and their rigidity. Conversely, red blood cells of individuals who are not sick are flexible, concave on both sides, and disk-shaped. Two hypotheses can explain the process by which the red blood cells take the sickle shape. The first is the cellular hypothesis: red blood cells' rigidity is explained by the cell membrane's properties. The second is molecular: this process is linked to chemical and physical properties of hemoglobin, a protein present in these cells. Pauling's team selected the molecular hypothesis and studied these chemical properties by comparing hemoglobin in anemic individuals, individuals with an anemic trait, and those lacking anomalous red blood cells. The notion of molecular disease thus comes partly from the distinction of three levels of analysis: clinical, cellular, and molecular. It also rests on the choice of explicative hypothesis of the molecular level. DNA is also a molecule, and it is easy enough to complete the schema by referring to an even more elementary level: genes that code proteins.

Yet to understand what signifies a molecular cause, we must be aware of the fact that researchers try to find the best explanation of sickle-cell disease's characteristic phenomenon. The best one is molecular: it is possible to distinguish two types of hemoglobin that differ by their electric charge, which also explains their difference in shape. Let's pause for a moment to go through this causal explanation. Hemoglobin, whose function is to carry oxygen in the organism, is described in biochemical terms, as a succession of amino acids (some of which are electrically charged). Because of the protein's electrical charge, each molecule of abnormal hemoglobin carries a region complementary to another hemoglobin molecule's region: these molecules can thus associate. The association of hemoglobin molecules leads to their alignment in the cell: the latter then takes the form of a sickle. This molecular explanation is both specific and complete.

It is but a short step from the concept of molecular disease to that of genetic disease. In 1957, Ingram analyzed amino acids comprising the two types of proteins and found the amino acid modified. He explains: "the sequence of base-pairs along the chain of nucleic acid provides the information which determines the sequence of amino-acids in the polypeptide chain for which the particular gene, or length of nucleic acid, is responsible. A substitution in the nucleic acid leads to a substitution in the polypeptide" (Ingram 1957, p. 328).

The terms "lead to" and "are responsible for" express a causal relationship, which is why researchers consider the DNA sequence to be the primary cause of the disease. In the case where an organism is the subject of study, it is effectively impossible to go any deeper than the DNA sequence to explain the abnormal protein's presence.

The example of the discovery of the "cystic fibrosis gene" also explains how disease comes to be seen as genetic. In 1985, the main symptoms of this disease were well known: chronic lung and respiratory infections and, in certain cases, insufficient pancreatic enzymes. Measuring chlorine levels in sweat allowed for diagnosis of the disease. Beginning in 1983, researchers knew that tissue of those with the disease was not permeable enough for chlorine; the disease could have perhaps been caused by a problem with chlorine ion transport. But they did not understand the "cause" of these symptoms and wanted to find a "basic defect" that caused the disease: despite intensive research effort, the basic defect in CF remains unidentified. Therefore, the metabolic abnormalities on which the biochemical studies are based are probably secondary or tertiary consequences of the primary defect (Tsui et al. 1985, p. 1054).

The direct strategy of studying the abnormal protein that Pauling used is impossible in the case of cystic fibrosis: the pathology's primary explanatory defect is unknown. The inverse genetic method allowed researchers to identify the cause of cystic fibrosis. Philip Kitcher describes this research strategy: "Even though biomedical researchers may initially be entirely ignorant about the physiological processes that go awry in a particular disease, knowing how that disease is transmitted in a sufficiently large sample of families, they can sometimes isolate the locus that is responsible. The strategy is to find genetic markers (...) associated with the transmission of the disease, confine the locus to a particular chromosomal region (...) pick out candidate genes, and ultimately, clone and sequence the desired gene. Knowledge of the gene may then yield enough understanding of the protein to provide insight into the causal basis of the disease" (Kitcher 1994, p. 522).

In the case of cystic fibrosis, the gene responsible for the pathology was identified and sequenced in 1989: it codes the CFTR (Cystic Fibrosis Transmembrane Conductance Regulator) protein, which is involved in ion transport.

The concept of a candidate gene is important: starting with a linkage study, a region of DNA is localized. Sequencing provides researchers with a long list of nitrogenous bases in which they can locate several groups that correspond to genes. The problem is thus one of knowing which, among all these possible genes, could be "the cystic fibrosis gene". Starting with the DNA sequence makes it possible to determine the protein's amino acid sequences and, in some cases, its function.

Researchers can therefore identify an order of priority as a function of structures and functions of the proteins that are coded by candidate genes. For some of these genes, it is plausible that their modification causes symptoms that are characteristic of the disease. The order of priority is thus defined based on physiological and pathological, rather than genetic, data.

The gene coding of the CFTR protein is selected in the following way (Riordan et al. 1989): we know that the conductivity of the chlorine ion across cell membranes diminishes in ill individuals; a good candidate gene would thus be a DNA sequence that codes a protein that contributes to the formation of an ionic channel or a DNA sequence that codes for a protein involved in the regulation of ionic channels. Second, we identify the RNA that corresponds to the DNA of CFTR in lung, colon, and sudoriparous tissues; this localization matches up with symptoms characteristic of the pathology. Finally, CFTR is analyzed: two groups of amino acids form a field capable of crossing a cell's plasma membrane; this data matches up with the hypothesis that the disease results from a dysfunctional ionic channel. In addition, certain amino acids of this protein are susceptible to form linking fields for molecules that intervene in regulation processes. And finally, the order of amino acids in this protein is compared to that of proteins in other species whose function is known. This comparison allows for the conclusion that the protein coded by the candidate gene is likely involved in ion transport back and forth in cells' plasma membrane. It is the coherence of this physiopathological data that gives meaning to the concept of genetic disease. Let's not forget one final observation: the reverse genetic study follows a genetic linkage study and is thus only possible if the method of disease transmission is known. What, then, is a genetic disease? It is a hereditary disease whose fundamental (molecular) defect has been identified at the level of DNA. This defect explains the disease's characteristics on higher levels (cellular, tissue, physiological).

The importance of a disease's hereditary dimension to its definition as a genetic disease can also be highlighted in the example of sickle cell disease. Ingram claims: "The latter [hemoglobin] is an abnormal protein which is inherited in a strictly Mendelian manner; it is now possible to show, for the first time, the effect of a single gene mutation as a change in one amino-acid of the hemoglobin polypeptide chain for the manufacture of which that gene is responsible" (Ingram 1957, p. 326).

It is precisely this identity between the gene as hereditary factor and the gene as DNA sequence that explains the importance of this research in elaborating the concept of genetic disease. Up until 1949, sickle-cell disease was considered to be a disease of dominant transmission and variable expressivity (variable expressivity referring to the fact that the disk-shaped red blood cells only took the sickle shape when oxygen pressure dropped). Geneticists assumed at the time that a single copy of a mutated gene was enough to explain the frequency of individuals in a family who were more or less anemic; the same allele of the gene was expressing itself in different ways. Neel, in a 1949 (Neel 1949) article, showed that sickle cell was transmitted recessively. His predecessors were mistaken because the clinical entity had not been well defined: the anemic trait and sickle-cell anemia were not sufficiently distinct from one another. Neel distinguished phenotypes (the anemic trait and sickle-cell anemia), drew up genealogical trees and observed the frequency of individuals who had and did not have the anemic trait and sickle-cell anemia. His calculations showed that the hypothesis of a recessive transmission had to be the better one. Pauling explained (Pauling et al. 1949) that he arrived at this same conclusion before Neel's article was published. The experiment with molecular biology in 1949 that indeed led to thinking that the mode of the disease's transmission was recessive; the cells of individuals with an anemic trait contained two types of hemoglobin, whereas healthy individuals' cells and anemic individuals' cells had only one type. Each type of protein could thus be translated from different alleles of the same gene. Research in classical genetics and molecular biology, even when carried out separately, confirmed this.

One objection does arise for the defense of genetic disease as a concept that relies on this double dimension of hereditary transmission and explanation at the molecular level of certain pathologies. What about complex genetic diseases whose heredity is not well known and that are only associated with specific risk factors?

Statistics Are Blind

Magnus's analysis also failed to account for the connection between statistical studies on genetic risk factors and molecular research that, as we have seen, are linked to physiopathology. It is clear why: statistical methods identify P-genes whose specific activities remain unknown at the molecular level. In the case of diseases that are transmitted following Mendel's laws, the "gene that causes the disease" thus had first and foremost an instrumental predictive value. Today, the Genome Wide Association Study (GWAS) is an association study. The goal is to show a difference in allelic frequency at the level of one single locus in individuals who are or are not affected by common diseases. As is the case for the P-gene, nothing is known about this allele's specific action on the pathology's development. Is Magnus right then to separate the explicative and statistical approaches?

Before I answer this question, it is important to point out a main connection between statistical studies and hereditary studies (which, incidentally, echoes the connection between hereditary and molecular dimensions to genetic diseases). To understand the genetic component of multifactorial diseases with complex heredity whose appearance depends on the interaction of several genes and of environmental factors, it is necessary to demonstrate their hereditary dimension. In their reference book, *Human Molecular Genetics*, Tom Strachan and Andrew Read (2004) separate the study of genetic diseases into two groups. One chapter, "Identifying human disease gene", is dedicated to methods that localize genes of diseases that are monogenic and Mendelian. Another chapter, "Mapping and identifying genes conferring susceptibility to complex diseases", deals with methods that identify genes for susceptibility and predisposition. Here, the way the authors present the difference between Mendelian and complex diseases is quite instructive. In effect, they explain that nobody would argue with the idea that a disease is genetic if it clearly follows a Mendelian mode of transmission, which assumes that a single gene is transmitted in a family. On the contrary, for complex traits, it is necessary *to prove that genetic factors are involved in the disease's development*. And for that to happen, research on the disease's transmission in genealogies is necessary. This type of research has, for example, led to the distinction between two different forms of Alzheimer's disease: a familial form (FAD or Familial Alzheimer's Disease) which is transmitted in a dominant autosomic manner, as well as a sporadic one, linked to APOE e4. Three predisposition genes (PSEN1, PSEN2 and APP) confer a very high risk of developing early-onset FAD. FAD is rare (5 %); the sporadic form is more frequent. There are also several ways of working with multifactorial diseases to calculate their genetic component (for a review, cf. Feingold 2005 and Campion 2001), but three steps are always required: "To show that the disease is familial, to show that this familial tendency is due to genetic factors, and, finally, to identify the genes involved" (Feingold 2005, p. 927). Research on genetic determinants thus rests on studies of the disease's mode of transmission in genealogies.

Finally, the most effective group of statistical methods is most meaningful when the research takes heredity into account. Familial aggregation studies attempt to find out the prevalence of a disease within relatives compared to their prevalence within the general population. Twin studies compare the similarity of monozygotic and dizygotic twins. Segregation studies collect genealogical trees and model the number of genes involved. For the association studies that do not assume any study of relatives but simply that of two populations (affected and not), a study of the affected subject and its two parents completes the analysis (Feingold 2005, p. 930). In a recent article in *Nature* (Manolio et al. 2009), the authors explain that GWAS have led to the identification of hundreds of genetic variations associated with diseases or complex human traits. However, these variants explain only a small proportion of heritability: it is thus necessary to find research strategies that go beyond GWAS to explain the rest.

The second connection Magnus misses concerns the links between statistical and explicative approaches to diseases. Association studies are often accompanied by research on candidate genes. Their selection is based on the coherence between their function and the characteristics of the pathology being studied. This makes sense: the difficulties that arise in the study of complex disease are linked to studies' statistical properties, for example problems with the threshold of significance or the population choice. They are also linked to the weakness of the relative risk conferred by each susceptibility allele. For each different pathology being studied, it is thus necessary to try to look elsewhere for what these alleles might modify in characteristic biological pathways.

Two examples serve to better explain this strategy. The first is Jean Dausset's article on histocompatibility systems and cancer risk (Dausset 1968). After Dausset showed in 1958 that the histocompatibility complex is hereditary, he wanted to explain why certain cancers, such as chronic lymphoid leukemias, are very frequent in some populations and nearly absent in others. When, in 1968, he discovered an association between antigenic differences characteristic of mouse populations and their resistance (or not) to a leukemia virus, Dausset immediately raised the following question: how to explain this association? In this study, the presence of specific alleles of the H-2 histocompatibility system was linked to a resistance to cancer; however, a link is not an explanation. "Resistance genes" (Dausset 1968, p. 1397) at locus H-2 or near it, might have explained this resistance, but no experiment "directly demonstrates the involvement of the H-2 locus itself" (ibid., p. 1398). Finally, according to the immunological explanation of the resistance phenomenon, the concept of cause in this context would designate a molecule that would prevent the virus from penetrating the cell. Demonstrating the role of the products of gene resistance in recognition phenomena of the antigen or of virus penetration in the cell was, at the time, beyond researchers' reach. This experiment only shows a correlation between certain alleles and a diminished frequency of virus contamination; it did not demonstrate causality. On the contrary, it showed that statistical research becomes meaningful in the context of attempts to explain genes' effects on pathological processes.

The second example is more contemporary. The difficulties encountered in statistical studies led researchers to progress toward showing evidence of more adequate relationships between more clearly described and individuated phenotypes and candidate genetic factors that may be causes. Dominique Campion reminds us that linkage analyses, which rely on knowing a disease's transmission mode, are most often inadaptable for the study of complex diseases. Association studies, for their part "are most interesting when it comes to candidate genes" (Campion 2001, p. 1139). A gene always intervenes in a biological pathway to whose modification it contributes. According to Campion, "for a long time, it was more or less implicitly understood that Mendelian and multigenic diseases involved two radically different biological characteristics" (ibid., p. 1144). In reality, in both cases genes intervene to modify biological pathways. In the case of monogenic disease, a single gene's mutation causes a major disruption in the pathway, which has important function consequences. In the case of multigenic diseases, the modifications are not as sever, but the effects of several involved risk factors on a single biological pathway can accumulate to the point where they cross a threshold that causes the pathology to appear. What then is the strategy for identifying good candidate genes? It is to first identify the biological pathways involved in a pathology's development.

This analysis clarifies why Magnus's "re-examination" of Huntington's disease is not as astonishing as he claims. This disease is linked to a characteristic repetition of CAG codons in a portion of chromosome 4. We know that it is difficult to predict whether individuals with between 30 and 40 repetitions will or will not be sick. This is a problem if we assume that the "Huntington's disease gene" works in an "all or nothing" way. But if the disease results from a disturbance in a biological pathway and if the effects of gene mutations are quantifiable, then it is easier to understand why the disease functions as a series of thresholds being reached and passed. The variable expressivity of monogenic diseases, such as symptoms' severity or the timing of their appearance, is also better explained by the threshold definition that corresponds to protein interactions with other variables involved in the biological pathway. Ultimately, it is not surprising that different genetic variants or even different genes have similar effects on biological pathways. For cystic fibrosis, there is an inventory of several different variants whose effects are more or less deleterious. Thalassemia is linked to mutations in several different genes.

In order to find the biological pathways that are disturbed in the context of genetic diseases, it is often necessary to identify Mendelian sub-entities and intermediary phenotypes are also often necessary. This is the case with diseases like schizophrenia, where difficulties in reproducing linkage and association studies have led researchers to identify intermediate phenotypes associated with a more readable genetic determinism. Campion (2001), for instance, points out difficulties with slow eye tracking and sensorial filtering, which are two endophenotypes associated with schizophrenia. Maziade's team (Maziade et al. 2003) refers to "dimensional" phenotypes (aggregates of symptoms) or neurocognitive phenotypes associated with the disease that are genetically less complex. Finally, there are genetic studies, which, in their difficulty, point researchers toward different nosographies: "schizophrenia" increasingly tends to be reduced to different etiologies while its borders with bipolar disorders become more blurred.

So, are we justified in considering multifactorial diseases as "genetic" diseases? It does not seem to be the case that researchers working on the genetics of complex diseases are trying to show that asthma or bipolar disorders are genetic diseases. Rather, they are using researching strategies that allow them to better understand the genetic factors involved in these diseases, since a better grasp of a disease's physiopathological development can lead to the identification of treatment options based on an improved understanding of interactions between genes and the environment. Genetic studies can even improve diagnoses: the HLA-B27 variant now helps diagnose ankylosing spondylitis.

The Choice of Genetics

I would like to return one last time to the issue of knowing why, among all the possible causes of a disease, the genetic ones are often privileged. Magnus ends his article with the following idea: epistemological answers to this question are insufficient. All that is left is to examine the use of the concept of genetic disease. He explains: "Labeling a disease as 'genetic' is to make an implicit claim that, for that disease, understanding and therapy will best come about through research at the genetic level. In other words smuggled into the very conceptual classification is a set of commitments about the best way to allocate resources and the best way to do good science and medicine" (Magnus 2004, p. 240).

Yet for Magnus, we are cruelly lacking in empiric arguments justifying the financing of genetic research. On the other hand, this remark demonstrates why it is necessary to identify values beyond the scientific ones that would lead to this type of research being privileged over others.

Looking at Kitcher's (2000) analysis of reasons that lead to genetic determinism's dominance in biology and medicine, despite our recognition of the environment's role in human development and behavior, serves to explain Magnus's argument in more depth. Kitcher's article is a response to Richard Lewontin, who "diagnoses errors that have seduced influential scholars and their readers into believing vulgar slogans about genes and destiny" (ibid., p. 283). Lewontin (1992) criticizes the popular conception of genetic determinism by pointing out that it ignores the interaction between genetic and non-genetic factors. This is why extending the category of genetic disease is socially dangerous: it leads us to think that what is "genetic" is "inevitable" and veers quite close to the belief that the only solution to "genetic" problems is the selection of individuals based on genetic criteria. For Lewontin, this popular conception belies a deeper concern: biology should, at least in part, be conceived of differently. Following Lewontin's reasoning Susan Oyama defends the argument that the oppositions biologists work on, for instance that between innate and acquired, or gene and environment, are not relevant (Oyama 1985). Focusing attention on the gene as a causal factor is simply an abstraction of complex causal situations that unfairly prioritizes certain determinants of the phenotype. Yet Kitcher aims to show that the interactionist concept, which separates genes from the environment, should not be rejected either. He argues: "Genetic determinism persists not because of some subtle error in conventional ideas about the general character of biological causation but because biologists who are studying complicated traits in complex organisms are prone to misapply correct general views" (Kitcher 2000, p. 284).

He is thus defending a "democratic" concept of environmental and genetic causes and explains why, in fact, scientists privilege genetic causes.

To demonstrate his argument Kitcher analyzes the reaction norm, a commonly used concept in biology. The underlying strategy in genetic determinist research on biological traits "begins by isolating certain properties of organisms for exploration of their causal impact, regarding the phenotype as the product of contributions from particular kinds of DNA sequences, on one hand, and from *everything else*, on the other. It goes on to inquire how the phenotype varies as the DNA sequences are held constant and as other factors (the cytoplasmic constitution of the zygote, the molecules passed across cell membranes, etc.) change" (ibid., p. 285).

The genotype's reaction norm is the graphic representation of this strategy: a phenotype's genetic determinism is defined as its relative invariance, given a single genotype in all environments. For opponents of the interactionist argument, the assumptions that are required to build the reaction norm, for instance the genotype-phenotype distinction, must be reconsidered. Kitcher, meanwhile, is trying to point out in his article that this tool is scientifically valid. Isolating certain causal factors by holding them constant in order to see how the effect varies when other factors are modified is legitimate.

But if this tool is scientifically valid, scientists often use it incorrectly. The interactionist effectively recognizes that there are several causes involved in development. Kitcher thus defends the principle of "causal democracy", where it is also scientifically justifiable to propose a causal analysis of a particular environmental factor by observing what happens when a genotype is modified: "The democracy principle accords no special privilege to the representations that foreground the role of genes" (Kitcher 2000, p. 290). Why, then, do biologists insist on the primacy of genetic causes?

The first reason Kitcher gives is that it is pragmatic: researchers believe that new technologies that come out of molecular biology can improve their understanding of certain diseases and certain behaviors. Kitcher explains that behavioral genetics promises much, because it must be possible to use DNA sequencing techniques to identify alleles shared among different individuals in a population. In this case, if the causally relevant environmental factors were identified (which poses the larger challenge to this type of study) it would be possible to study phenotypes' variation as a function of a genotype held constant when environmental factors change. In the medical field, he uses the example of research on alcoholism and addiction, and explains why scientists are attempting to identify genetic causes: "They begin with genetic causes not because they are convinced that these are the most important (that the norms of reaction for certain 'addictive' genotypes are virtually flat) but because they want to unravel the neurochemistry, and they see the investigation of genotypes as a thread that will lead them into the tangle" (ibid., p. 295).

And, in a sense, this is what is happening, since this type of strategy identifies multiple nosological entities for a pathology that was previously considered singular. Modes of transmission for these traits in families and eventual genetic interactions are better understood. Our understanding of pathologies improves. In the case Kitcher discusses, researchers hope that the inverse genetic method and knowledge of the DNA sequence will lead to an understanding of molecular changes in the brain. These scientists believe that they understand the molecular details of interactions carried out between the organism and the environment, which differ between addicted and non-addicted people. Here, then, it is unnecessary to assume a social deformation of science that would rely on socially valued norms to justify the importance granted to research on genetic determinants of diseases.

Yet according to Kitcher, we would assume that a study of the reaction norms of "violence alleles" would be socially encouraged for reasons that are not scientific or medical. Under the principle of causal democracy, it would be possible to carry out a study in which the environment would be held constant and the variation of phenotypes would be observed as a function of this causal factor. In this case, according to Kitcher, we could see that for a single genotype, the phenotype varies based on environment and we could thus conclude that the environment is a causal factor in violent behavior. Kitcher, however, writes: "In a society that consistently and callously turns its back on programs that might aid the unfortunate and sees taxation as a form of robbery rather than a necessary means to social cooperation, the investigation I have outlined has no obvious point" (ibid., p. 296).

On the contrary, studies that hope to show that keeping a genotype constant and varying environments leads to an invariable phenotype (the reaction norm is flat or nearly flat) have a greater chance of finding funding. In this case, detecting this genotype in individuals would identify a predisposition to violent behavior. This type of study would thus reinforce the idea that social solutions are hopeless: if an individual is predisposed to violence, there is little chance that school, for instance, would help him or her escape this fate. Genetics are considered inevitable, despite the fact that all scientists agree that genes alone determine no trait.

This way of thinking has been subject to numerous critiques since the 1990s. The concept of "geneticization" arose as a criticism of the extension of a genetic vision of man accompanied by scientific discoveries. Abby Lippman (1991) and Henk Ten Have (2001) view it as a process that affects medicine as well as the broader society. Geneticization redefines individuals according to their genes and creates a new language to describe, interpret and understand human life: it explains human differences by their genetic differences. This trend is deterministic and reductive, and it bears serious consequences for the definition of genetic disease.

To begin with, in medicine, extending the category of genetic disease empties the term of its specificity; it would cease to carry any nosographic meaning. Thus, if the notion of genetic terrain is to have meaning, it must not lead to every disease being "genetic" under the pretext that some alleles are associated with a greater risk for a given disease.

Second, geneticization has consequences for the instrumental approach to cause. Stating that X is genetic is, according to the critique of geneticization, to abandon any public policy that tries to improve environmental conditions, since "genetic" here is the equivalent of "inevitable". At the same time, considering a disease genetic is to also believe that the easiest course of treatment lies at the genetic level. In fact, the most common type of management of genetic diseases is fetal screening with genetic tests. The possible consequences of such processes are indeed worrisome. The extension of the concept of genetic disease effectively leads to practices that raise the specter of eugenics. In current debates, the slippery slope argument is frequently raised to underscore the danger of extending prenatal or preimplantation diagnostic practices in the context of genetic predisposition. For example, authorizing such diagnostic testing for diseases that could manifest themselves later on and whose likelihood of appearing is not equal to 1, such as hereditary ovarian and breast cancers, could open the door for birth selections based on much more trivial criteria. From this perspective, there is much at stake in the definition of genetic disease.

Conclusion

Three criteria thus function together to define a disease as genetic. The first concerns specific causality that allows for the understanding of the pathology's development. A gene is in this context an explicatory unit for a specific modification of an important biological pathway within a pathology's framework. The second criterion is the link between these explanations and the hereditary dimension of diseases, approached using the framework of statistical studies. The last criterion is the perspective of preventive or curative medical care that is compatible with ethical imperatives. The articulation of these criteria could be extended to many more timely reflections on the concept of disease. For example, homosexuality was removed from the DSM (Diagnostic and Statistical Manual of Mental Disorders) in

1973. Consider this: if homosexual practices had been correlated to a group of genetic risk factors at the time, would they still have been pulled from the list of mental disorders?

References

Campion D (2001) Dissection génétique des maladies à hérédité complexe. Médecine/sciences 17:1139–1148

Dausset J (1968) Les systèmes d'histocompatibilité et la susceptibilité au cancer. Presse Med 76(28):1397–1400

Feingold J (2005) Maladies multifactorielles un cauchemar pour le généticien. Médecine/sciences 11(21):927–933

Feldman D, Tauber A (1997) Sickle cell anemia: reexamining the first molecular disease. Bull Hist Med 71(4):623–650

Gannett L (1999) What's in a cause? The pragmatic dimensions of genetic explanations. Biol Philos 14:349–374

Gifford F (1990) Genetic traits. Biol Philos 5(3):327–347

Hesslow G (1984) What is a genetic disease? On the relative importance of causes. In: Nordenfelt L, Lindahl BIB (eds) Health, disease and causal explanation in medicine. Reidel, Dordrecht, pp 183–193

Hull R (1979) Why genetic disease? In: Capron AM et al (eds) Genetic counseling facts, values and norms. Alan R. Liss, New York, pp 57–69

Ingram VM (1957) Gene mutations in human haemoglobin: the chemical difference between normal and sickle cell haemoglobin. Nature 180:326–328

Ingram VM, Stretton AO (1959) Genetic basis of the thalassemia diseases. Nature 184:1903–1909

Kitcher P (1994) Who's afraid of the human genome project? In: Hull DL, Ruse M (eds) The philosophy of biology. Oxford University Press, Oxford, 1998, pp 522–535

Kitcher P (2000) Battling the undead. How (and how not) to resist genetic determinism. In: Kitcher P (ed) In Mendel's mirror. Philosophical reflections on biology. Oxford University Press, Oxford, 2003, pp 283–300

Kitcher P (2003) In Mendel's mirror: philosophical reflections on biology. Oxford University Press, Oxford

Lewontin R (1992) Biology as ideology. The doctrine of DNA. Richard Harper Perennial, New York

Lippman A (1991) Prenatal genetic testing and screening: constructing needs and reinforcing inequities. Am J Law Med 17:15–50

Magnus D (2004) The concept of genetic disease. In: Caplan A, McCartney J, Sisti D (eds) Health, disease, and illness. Georgetown University Press, Washington, DC, pp 233–242

Manolio TA, Collins FS, Cox NJ, Goldstein DB, Hindorff LA (2009) Finding the missing heritability of complex diseases. Nature 461:747–753

Maziade M, Merette M, Chagnon Y-C, Roy M-C (2003) Génétique de la schizophrénie et de la maladie bipolaire. Médecine/sciences 19(10):960–966

Moss L (2004) What genes can't do. MIT Press, Cambridge, MA

Neel JV (1949) The inheritance of sickle cell anemia. Science 110:64–66

Oyama S (1985) The ontogeny of information. Developmental system and evolution. Cambridge University Press, Cambridge

Pauling L, Itano HA, Singer SJ, Wells IC (1949) Sickle cell anemia, a molecular disease. Science 110:543–548

Riordan JR, Rommens JM, Kerem B, Alon N, Rozmahel R, Grzelcak Z (1989) Identification of the cystic fibrosis gene: cloning and characterization of complementary DNA. Science 245:1066–1073

Sober E (2001) The meaning of genetic causation. In: Buchanan A, Brock D, Daniels N, Wikler D (eds) From chance to choice. Genetics and justice. Cambridge University Press, Cambridge, pp 347–370

Sterelny K, Kitcher P (1988) The return of the gene. J Philos 85:339–361

Strachan T, Read AP (2004) Human molecular genetics 3. Garland Science, New York

Ten Have HAMJ (2001) Genetics and culture: the geneticization thesis. Med Health Care Philos 4:295–304

Tsui L-C, Buchwald M, Barker D, Braman JC, Knowlton R (1985) Cystic fibrosis locus defined by a genetically linked polymorphic DNA marker. Science 230:1054–1057

Causal and Probabilistic Inferences in Diagnostic Reasoning: Historical Insight into the Contemporary Debate

Joël Coste

Abstract This paper first reviews contemporary incarnations of probabilistic, causal and pathophysiological reasoning and their efficacy and limitations in medical diagnostic practice. It is then shown that diagnostic and causal reasoning has been closely associated throughout the history of rational medicine in a relationship that may be seen as consubstantial, or even central to the very foundations of rational medicine. This relationship both explains the long-standing and persistent nature of the tensions between the "deterministic" approaches used in medical practice, particularly in diagnosis, and the early emergence of alternative or so-called "empirical" approaches, which have been maintained right up to the present day when they are expressed in the form of clinical decision rules.

> We suggest that applying a probabilistic technique can considerably improve the precision of tissue diagnosis and can greatly facilitate the communication of pathologists with clinicians and with each other. Probabilistic analysis is also likely to be of substantial value in improving the interpretation and reporting of x-ray and nuclear-medicine studies.[1]

> Medical knowledge is growing at an explosive rate. While the availability of pertinent data has the potential to make the task of diagnosis more accurate, it is also increasingly overwhelming for physicians to assimilate. [...] Bayesian networks have the potential to provide decision support in radiology because they can model uncertainty, calculate and explain post-test probabilities, and integrate a large amount of information efficiently.[2]

For more than 30 years now, the medical press has regularly announced that only probabilistic approaches *have the potential* to manage the "explosive" increase in diagnostic knowledge. Whatever the actual state of this knowledge—currently being increased constantly by data from new "complementary examinations", which

---

[1] Schwartz et al. (1981).

[2] Burnside (2003).

J. Coste (✉)
EPHE (Sciences Historiques et Philologiques), Université Paris Descartes, Paris, France
e-mail: joel.coste@htd.aphp.fr

P. Huneman et al. (eds.), *Classification, Disease and Evidence*, History,
Philosophy and Theory of the Life Sciences 7, DOI 10.1007/978-94-017-8887-8_8,
© Springer Science+Business Media Dordrecht 2015

are often ephemeral products of a certain type of technological and commercial medicine—the almost invariable association of probabilistic calculation with the word "could" and the idea of "potential" suggests a certain *resistance* to this probabilistic calculation in the field of practical medicine. Taking a step back from the enthralling and often passionate, but also highly technical, debate on probabilistic reasoning that has been agitating the medical community for more than three decades,[3] our aim here will be to analyse one of the reasons for the resistance to the use of Bayesian arithmetic: the existence of powerful alternative types of medical reasoning. These alternatives, causal and pathophysiological reasoning, which are based on the *causes* and *mechanisms* of disease, respectively, have been seen for centuries as well-suited to the reality of diagnostic practice. We will first briefly present contemporary incarnations of these three types of diagnostic reasoning and examine their efficacy and limitations in current practice. We will then recall some of the history of medicine to show that diagnostic and causal reasoning have been closely associated throughout the history of learned medicine,[4] in a relationship that may be seen as consubstantial or even as central to the very foundations of learned medicine. The history of medicine will also remind us that since the (Hippocratic) origins of learned medicine, there have been tensions concerning the "deterministic" reasoning and approaches to medical practice, particularly for diagnosis. Indeed, these tensions probably contributed to the development of so-called "empirical" approaches, which are not unrelated to the contemporary Evidence-Based Medicine approach of diagnosis. We will, of course, focus on only a few moments in the history of medicine and a few aspects of the history of medical diagnosis that remains largely unknown to historians and epistemologists of medicine, despite the fact that they may usefully enlighten contemporary debates about diagnosis.

Contemporary Incarnations of Probabilistic, Causal and Pathophysiological Reasoning for Medical Diagnosis[5]

Diagnosis is classically defined as the art of recognizing diseases from their symptoms and of distinguishing between them. The *recognition* of diseases is based on the comparison of symptoms and signs observed in a given patient with the reference symptoms and signs comprising the *conceptual representation* of the pathological entity with which the doctor believes that he or she is faced. *For this*

---

[3] For an example of early debates on probabilistic reasoning, see Ransohoff and Feinstein (1976).

[4] In this paper, we use the expression "learned medicine" to refer to medicine taught through a regular and often lengthy process of education, conducted in medical schools or colleges attached to universities since the thirteenth century in Western Europe.

[5] This first part of the article is based largely on chapters 4, 10 and 13 of the book cowritten with Jean-Baptiste Paolaggi (1928–2010), Le raisonnement médical. De la science à la pratique clinique (*Medical reasoning. From science to clinical practice*), Editions ESTEM, Paris, 2001.

recognition, the doctor must, of course, know and recall nosological information about conditions (for the important issue of nosological constructions, see below). Diagnosis may be *extremely simple*, because the nature of the condition concerned is simple or because the doctor is highly experienced. Some diseases, such as boils or open fractures of major bones, may even be diagnosed by laypeople. However, in most cases, diagnosis is a gradual process, which advances as information is collected during a structured observation of the patient through the use of *reasoning*, making it possible to advance to a point at which the doctor believes he or she has found a satisfactory degree of agreement between the observed signs and those of a defined pathological entity. (It may not be possible to obtain a satisfactory diagnosis straight away due to the lack of symptoms at a certain stage of disease progression or because additional information is required: In such cases, doctors must content themselves with taking the necessary measures to save the patient's life or functions whilst awaiting the possibility of making more satisfactory progress).

The Principal Types of Reasoning Used in Diagnostic Approaches

Doctors use diverse types of reasoning in their diagnostic approaches, and they may even use several types of reasoning together for an individual case. Descriptive, causal, pathophysiological and probabilistic reasonings constitute a group of reasoning widely accepted within the scientific and medical community. A second group is more controversial: analogical reasoning and reasoning based on or supported by algorithms or *clinical prediction rules*.[6]

Probabilistic reasoning is based on the use of probability calculations to determine the "posterior probability" of a disease (D) or, in other words, the probability of the disease given the results of a diagnostic test. The probabilities P(D+/T+), if the test is positive, and P(D+/T−), if the test is negative, can be calculated from Bayes' theorem using a "prior" probability of the disease P(D+) and the metrological characteristics of the test: its sensitivity (Se), defined as the probability of a positive test result in a subject with the disease, and specificity (Sp), defined as the probability of a negative test result in a subject without the disease.

$$P(D+/T+) = \frac{Se \times P(D+)}{Se \times P(D+) + (1 - Sp) \times (1 - P(D+))}$$

---

[6] In addition to these explicit and conscious reasonings, in clinical practice doctors also frequently make use of types of *subreasoning* described as "heuristics", including, in particular the representativeness heuristic (closely resembling shape recognition), the availability heuristic and the anchoring and adjustment heuristic (which are actually modes of implicit probability management). For further information about heuristics beyond the scope of this article, please see, in particular, Tversky and Kahneman (1974).

$$P(D+/T-) = \frac{(1-Se) \times P(D+)}{(1-Se) \times P(D+) + Sp \times (1-P(D+))}$$

Despite its mathematical formulation and its simplicity, probabilistic reasoning is very rarely used *directly* in everyday medical practice. The material difficulties involved in its implementation (with the need to use a computer or a calculator) and insufficient knowledge of even the most elementary metrological characteristics of the tests used, are generally blamed for the underuse of these methods. By contrast, probabilistic reasoning has, over the last 15 years or so, been used *indirectly*, through the implementation of *clinical prediction rules* derived from the modeling of clinical situations with probabilistic models. These rules make it possible to retain or reject a diagnostic hypothesis on the basis of a small number of criteria or the value of a score constructed with these criteria. For example, the Ottawa rule can be used to eliminate the possibility of an ankle fracture following an accident, if the subject is able to stand on one leg (the leg on the injured side) and if palpation of the two malleolar regions does not cause intense pain.[7] The use of clinical prediction rules has been encouraged, in particular, by supporters of Evidence-Based Medicine, who see them as "convenient and rapid way(s) to apply the results of research to the care of patients and to help physician(s) make more accurate decision(s)".[8]

Causal reasoning is principally used for aetiological diagnosis, which is the attribution of an observed syndrome to a cause identified by medical science. This cause may be an infection, a tumour, a vascular problem, traumatic or inflammatory, for example. Its other main use is for identifying curative treatment "indications", a logical extension of the preceding approach (action against the cause logically leading to the abolition or attenuation of the pathological effect). Causal reasoning in the strict sense of the term considers the causes of morbid phenomena without taking into account the detailed mechanisms underlying the problem. It therefore remains upstream from pathophysiological knowledge and this can be problematic. For example, a microbe may cause an infection, but the problems induced by the penetration and development of this microbe in the body may differ considerably between subjects, from visceral problems in one subject to immunological problems in another. Not taking pathophysiological mechanisms into account in such situations may have harmful consequences for the patient. A good example is provided by toxoplasmosis, the signs of which differ considerably between children, normal adults, pregnant women and immunocompromised subjects. Furthermore, causal reasoning is clearly inappropriate for (many) conditions that are uncontestable in terms of their symptoms, with well-defined lesions, for which the physiopathological events have been partly resolved, but for which the exact cause remains unknown.

---

[7] Stiell et al. (1994).

[8] Ebell (2001, p. 2). For further information on this approach to learning and medical practice, first presented in 1992, see J. Coste and J.B. Paolaggi, Le raisonnement médical, op. cit., pp. 141–145.

Pathophysiological reasoning encompasses the use of *combined* knowledge from clinical and fundamental disciplines to obtain insight into the mechanisms underlying pathological problems. Information could be and has been used, successively and cumulatively, from the following fundamental disciplines: functional physiology, experimental medicine, bacteriology, cell physiology, biophysics, genetics and molecular biology. Pathophysiological reasoning is widely used in diagnostics, and is highly effective in many situations. In neurology, the site of nervous system lesions can be deduced from observations of the problems experienced by patients and their connection with functions of the territories and nuclei of the nervous system and the trajectories of the various nerves. For example, walking difficulties associated with a leg motor deficit with strong, diffuse reflexes, Babinski's sign (indicative of a "pyramidal syndrome", named after the motor nerve pathways), a lack of sensitivity to pain in the other leg and several sphincter problems, are immediately suggestive of damage to the dorsal or lumbar spinal cord. Similarly, a diagnosis of anaemia may, once recognised, lead, through reasoning concerning the mechanisms involved, to consideration of the size of the red blood cells, serum iron and ferritin concentrations and associated signs of haemolysis. In endocrinology, an understanding of feedback control may immediately lead to investigations of central, hypothalamic/pituitary or peripheral origins for hyper/hypothyroidism and hyper/hypocorticism. Pathophysiological reasoning has also provided the key for the aetiological diagnosis of rare diseases such as vitamin-resistant rickets and osteomalacia. Similarly, pathophysiological reasoning (as implemented initially by Garrod and subsequently by Ball[9]) led to the implication of lead and chronic saturnism in the "epidemic" of gout in port drinkers in England in the nineteenth century (port was fortified with lead for its transport) and in illegal whiskey distillers in the United States in the 1960s. In these situations, the strong hyperuricaemia responsible for the symptoms of gout was explained by kidney damage (toxic tubular/interstitial nephropathy) leading to a decrease in the urinary excretion of uric acid, and by the disruption of purine metabolism due to the toxic effects of lead, leading to a secondary increase in uric acid levels. However, alongside brilliant successes of this type, pathophysiological reasoning has often been let down by confusion between hypotheses and demonstrated facts in biology. It is sometimes difficult, in the context of biological medical research, to demonstrate a pathophysiological hypothesis rigorously and unequivocally. This has led to some biologists displaying excessive enthusiasm for their supposed discoveries, thereby weakening the nosological constructions on which the reasoning of doctors is based.

Contemporary Nosological Entities and Constructions

In our book *Le raisonnement médical, de la science à la pratique clinique* (Medical reasoning, from science to clinical practice), we suggested that several groups of nosological entities or constructions could be distinguished (Table 1) on the basis of

---

[9] Garrod (1859) and Ball (1971).

Table 1 Type of nosological entity as a function of signs, lesions, mechanisms and cause

Group 6	Group 5	Group 4	Group 3	Group 2	Group 1
Condition defined by: **Single disorder**	Condition defined by: **Several signs or a syndrome**	Condition defined by: **Several signs or a syndrome**	Condition defined by: **Signs or a syndrome**	Condition defined by: **Signs or a syndrome**	Condition due to an: **Innate DNA abnormality**
		+ **Progression**	+ **Progression**	+ **Progression**	+ **Persistence**
			+ **Lesions** or **Pathophysiology**	+ **Lesions** or **Pathophysiology**	+ **Clinical, lesional and pathophysiological consequences**
				+ **Cause**	

the validation of their defining elements, including, in particular, their cause and pathophysiological mechanism:

– The first group consists of entirely defined diseases, such as monogenic conditions for which the entire sequence of events, from DNA mutation to symptoms and their consequences for progression, has been determined (e.g. cystic fibrosis);
– The second group consists essentially of infectious diseases caused by identified agents, which may be bacterial (e.g. typhoid), viral (e.g. HIV) or parasitic (e.g. malaria) in nature. Some cancers with demonstrated viral causes (e.g. cervical cancer) caused by certain types of papillomavirus also belong to this group. For these conditions, a causal mechanism and a precise pathophysiology have been established;
– Other cancers (in which cell proliferation itself underlies the disease) form the third group, together with certain haematological, endocrinological and metabolic conditions and ischemic tissue necrosis. In this group, the pathophysiological mechanisms are well identified but the causes triggering these diseases remain unclear;
– The fourth group consists of diseases with known symptoms and progression and, for some, lesions, but for which the fundamental nature of, and precise reasons triggering, the condition remain unknown. The diseases in this group include multiple sclerosis and rheumatoid arthritis, for which no one doubts the existence and solidity of the specification, but for which the cause and underlying mechanisms are currently the subject of suppositions or highly imperfect demonstrations;
– The last two groups consist of diseases defined solely on the basis of their symptoms. Indeed, some conditions can be defined only by an association of symptoms or as a "syndrome" and the lesions may not be characterised with any certainty (as in lumbago, for example). In some cases, even the symptoms may be unclear. This is the case for "idiopathic" diseases, such as irritable bowel syndrome and many mental illnesses for which pathophysiological studies have yet to identify the mechanism and for which definitions are regularly revised.

This brief overview of current nosological entities and constructions explains why causal and pathophysiological reasoning methods are relevant only for entities with known causes and mechanisms, respectively. Such entities are not very numerous, resulting in perfectly justified criticism of the application of these types of reasoning in many situations in which probabilistic reasoning, or the use of clinical predication rules, if such rules exist and have been effectively validated, would clearly have been preferable.

Key Role of Causal Inferences in the Diagnostic Approach Throughout the History of Learned Medicine

What we have just said about the objective of diagnosis being to recognise *the* disease affecting *the* patient, and its contemporary implementation, might suggest that the idea of diagnosis has always been evident to medicine and to doctors. In fact, nothing could be further from the truth.

A study of terminology and its uses is useful here, as in many cases. The word "diagnosis" comes from the Greek διάγνωση (diagnosis: "to discern", *dignosco* in Latin) and was not widely used in a medical context before the sixteenth century, and did not become commonplace until the eighteenth century. Before this, a related, but not entirely equivalent, term/notion was used: "to differentiate" (διαφοροποιώ in Greek). For instance, Galen differentiated, that is, classified or *divided* symptoms, diseases and their causes according to the well known "Porphyrian tree" model. For example, diseases were divided into simple and complex diseases; simple diseases were divided into inflammations and distempers and distempers were divided into hot and cold. For the analysis of clinical cases, Galen put on an equal footing diseases (which, for him, resulted from a "diathesis", a condition affecting the entire body, or a "lesion" of a function or organ); symptoms (signs that appear abnormal to the patient or the doctor, not necessarily attributable to a disease—Galen distinguished two types: common and specific); and "causes", of both symptoms and diseases, considered one by one, in decreasing order of precision. This conceptualisation may appear somewhat confused, particularly as the writings of Galen are sometimes contradictory, but it becomes clearer when we recall that Galen drew largely on the *dynamic* Hippocratic concept of disease, in which the disease is not considered *per se*, but only in the context of *the disease in the patient*: considerable emphasis is placed on analysis of the imbalance responsible for the problem, taking into account the (present) pathological state of the subjects, their (previous) state of balance and the reasons for the loss of this balance. For the doctor, it is more a question of *knowing* than of *recognising* (diagnosing) the disease, through a multidimensional evaluation aiming to establish "what is wrong" and to determine how to re-establish balance. In this evaluation, the analysis of causes was clearly central to the therapeutic act, which was designed to return the subject to equilibrium, generally by effects opposing the causes giving rise to the pathological state. This multidimensional analysis of patients, strange as it may seem at the start of the twenty-first century, was not entirely without relevance, because even today it must be borne in mind that many symptoms are not linked to a disease, many are linked to several diseases and some display a continuum between a disease state and good health. The *dynamic* concept of disease, also described as "idiosyncratic" because there is no disease *per se*, just *a disease in a patient*, was prevalent in Western medicine until at least the sixteenth century; this is illustrated by the works of Jacques Dubois and Jean Fernel, for example, who suggested that doctors confronted with a given pathological situation should differentiate, classify or divide the symptoms, diseases and *causes*, before implementing any treatment.[10]

---

[10] Jacques Dubois tried, between 1530 and 1540, to summarise Galenic concepts in his work *Methodus sex librorum Galeni in differentiis et causis morborum et symptomatum in tabellas sex ordine...* followed by *De signis omnibus medicis hoc est, salubribus, insalubribus, et neutris, commentarius omnino necessarius medico futuro*. For the benefit of students, he constructed "division tables" (subforms of Porphyrian trees) presenting all the "differences" (types) of diseases and causes, and the differences and causes of symptoms, based on *De differentiis et causis morborum et symptomatum*; and tables of all the signs of health, neutral states and disease, based on *Ars medica* (principally), *On the Temperaments*, *On Affected Places* and on *Commentary to*

The focus on the fundamental need to *distinguish between diseases* responsible for pathological problems, to "diagnose" in the current sense of the term, began to become important with the development of *ontological* concepts of disease, in which diseases are seen as entities *independently* of the patient, in the sixteenth and, particularly, seventeenth centuries. This movement actually represented a *return* to ontological concepts in medical thinking rather than a *new* development.[11] There were premises for this return from the middle of the sixteenth century, in the work of Fernel, but particularly in that of Fracastor and van Lom (or Lommius,[12] a Dutch author who was undoubtedly influenced by Celsus, a first century Roman author, whose writings on this subject were original). Ontological concepts are also found in the theories of chemists, but they were not fully expressed in the context of medicine until Thomas Sydenham at the end of the seventeenth century. Sydenham unambiguously outlined, in the introduction of his *Observationes Medica*, the "species of disease" and the need to distinguish clearly between them, in the way that botanists distinguish between plants, *to ensure that patients receive the most appropriate treatment*:

> In the first place, it is necessary that all diseases be reduced to definite and certain species, and that, with the same care which we see exhibited by botanists in their phytologies; since it happens, at present, that many diseases, although included in the same genus, mentioned with a common nomenclature, and resembling one another in several symptoms, are, notwithstanding, different in their natures, and require a different medical treatment.[13]

Sydenham's work clearly marked a turning point in the conceptualisation of diseases and of diagnosis. It also marked a turning point in nosological research, which boomed in the eighteenth century culminating in the well known work of Boissier de Sauvages, published in 1732 and 1771. The very title of this work,

---

Hippocrates' On Prognosis. Dubois devoted only a little space to *On Affected places* and he did not mention the method for localising problems in chapter 5 of Book 1, unlike Argentorio (*De morbis* 1558) and, above all, Fernel (*Pathologia*, 1554), who attributed to this method an essential role in *knowledge* of the disease affecting the patient. "Fernelian diagnosis" involved *two stages*: the first involved "identifying the place of the illness", by following a method very similar to that outlined in chapter 5 of book 1 of *Affected places* and then "recognising the disease and its cause". The signs used to identify the affected place included "excrements", the features of the "lesion of function", those of the pain and of the "specific accidents" and were also subsequently used to identify the disease and its cause.

[11] For this point, see the analyses of Mirko Grmek, particularly in Volume 2 of "The History of Medical Thought" (Paris, Seuil 1997, p. 157 sq.).

[12] According to Van Lom, identification of the type of disease was a prerequisite for its correct management. *Medicinalium observationum libri tres, quibus notae morborum omnium et quae de his possint haberi praesagia judiciaque roponuntur* (1560) included a first part devoted to the "recognition" (van Lom used the verb "animadvertere") of general diseases (essentially fevers), a second part devoted to the recognition of diseases specific to parts of the body (classified from the head to the feet, and then by pathological process: inflammation, gangrene, erysipelas etc.) and a third part dedicated to prognostic signs. For more information about this work, see J. Duffin (2006).

[13] *Observationes medica*, London, 1676 (Translated by W.A. Greenhill, The works of Thomas Sydenham, Vol. 1, London, The Sydenham Society, 1848, p. 13).

Nosologia methodica, sistens morborum classes, genera et species juxta Sydenhami mentem et botanicorum ordinem, is particularly enlightening as concerns the debt that the author felt he owed Sydenham. However, whereas the nosology of Boissier de Sauvage's work was voluntarily descriptive and "botanical" in nature, with the causes of diseases clearly taking a back seat, these causes came to the forefront in the *Nosographie philosophique* of Pinel, which was published in 1798. Pinel insisted on the need to *order diseases by "laws"*, classifying diseases as fevers, phlegmasias, haemorrhages, neuroses, lymph system diseases and diseases of unknown location.[14]

This mention of Pinel allows us to make the transition to "classical" clinical diagnosis, which became established during the course of the nineteenth century and which has been taught, if not practised, ever since, in the form of the famous triad "positive diagnosis, aetiological diagnosis, differential diagnosis":

- Positive or syndromic diagnosis involves recognising the syndrome—a collection of symptoms and signs constituting an entity that can be recognised on the basis of the uniformity of the stereotypical combination of morbid signs or its effects on an organ or well-defined system—such as meningeal syndrome, febrile diarrhoea syndrome, acute arthritis, cerebellar syndrome or pyramidal syndrome.
- Aetiological diagnosis involves identifying the probable cause of the syndrome in the patient. A broad range of possible causes is often considered: traumatism, toxicity, nutrition, infection, vascular causes, tumours, immune dysregulation, degeneration, genetic, congenital or idiopathic causes.[15]
- Differential diagnosis involves the elimination of similar conditions. If we return to the example cited above (a subject with difficulty walking associated with leg motor deficit, in whom a pyramidal syndrome is found on the ipsilateral side, with impaired pain perception on the contralateral side and some sphincter problems), the symptoms observed strongly suggest a diagnosis of medullary syndrome. Aetiological diagnosis would involve searching for mechanical compression or compression due to a tumour, a vascular cause (e.g. angioma) or a malformation (e.g. syringomyelia). Differential diagnosis would involve the elimination of similar conditions, such as peripheral nerve compression (absence of pyramidal syndrome), polyradiculoneuritis (which is usually bilateral), multiple sclerosis (which is usually associated with other signs and a history of regressive deficits) and conversion disorder (but the systematisation of these problems is not usual).

Like Galenic diagnosis or, more accurately, evaluation, "classical" clinical diagnosis focuses principally on a search for causes. Throughout almost the entire

---

[14] Importantly, since Pinel, all nosologies and disease classifications, right up to the most recent revisions of the international classification of diseases (ICD 9 and 10), can be seen as mixed, or even assorted, with symptoms, syndromes and diseases of known or unknown cause.

[15] "Congenital", "degenerative" and "idiopathic" variants cannot, of course, be placed on the same plane as other "causes" in this operational categorisation for clinical practice.

history of learned medical practice, this search has been given precedence over other aspects, *paradoxically* in some ways, because the causes of diseases are generally not precisely known.

However, it should be stressed that causality is broad, complex and multiple, and that it operates at several levels[16]: the distant causes of diseases are often considered unattainable, but the underlying mechanisms, referred to today as "pathophysiological" mechanisms, are accessible and can be used for diagnostic reasoning, at least that was what was thought.

The Founding Causal Preoccupations of Learned Medical Practice

The key position of the search for causes in the "diagnostic approach"[17] is clearly apparent throughout the history of learned medicine. We will end this brief study by recalling that the causal preoccupations of this approach were the elements forming the very foundations of learned medicine. The major texts of the Hippocratic Corpus and those of Galen strongly highlight the importance of this search for causes— natural causes of course—responsible for the problems presented by patients *with a view to improving patient management*. Two Hippocratic texts, *Breaths* and *On Ancient Medicine*, the second being thought to have been written by Hippocrates himself, are of particular importance:

> Now of these obscure matters one is the cause of diseases, what the beginning and source is whence come affections of the body. For knowledge of the cause of a disease will enable one to administer to the body what things are advantageous. Indeed this sort of medicine is quite natural.[18]
>
> I think a physician must know, and be at great pains to know, about natural science, if he is going to perform aught of his duty, what man is in relation to foods and drinks, and to habits generally, and what will be the effects of each on each individual. It is not sufficient to learn simply that cheese is a bad food, as it gives a pain to one who eats a surfeit of it; we must know what the pain is, the reasons for it, and which constituent of man is harmfully affected. For there are many other bad foods and bad drinks, which affect a man in different ways, I would therefore have the point put thus: "Undiluted wine, drunk in large quantity,

---

[16] The four causes of Aristotle ("material causes", of which things were made; "formal causes", the idea of things or of models according to which they were made; "efficient causes", which made things and "final causes", the final or completed state of things) were widely used in Mediaeval medicine and at the start of the modern era. However, the three categories of causes in the Galenic system are the most extensively used by doctors: "antecedent causes", corresponding schematically to internal factors predisposing a subject to disease, "initial causes", which are generally external and trigger the disease and "cohesive causes", which are related to the pathophysiological process itself.

[17] We are of course dealing here with the diagnostic approach in the broad sense of the term, as described in the previous section.

[18] Hippocrates, Breaths 1 (translated by W. H. S. Jones, Cambridge, MA: Harvard University Press, The Loeb Classical Library Vol. 2, 1923, pp. 227–229).

produces a certain effect upon a man." All who know this would realise that this is a power of wine, and that wine itself is to blame, and we know through what parts of a man it chiefly exerts this power. Such nicety of truth I wish to be manifest in all other instances.[19]

This demand for the identification of causes nevertheless created very strong tensions and ill ease among the promoters of learned medicine, who were well aware of the difficulty, or even impossibility, of identifying the cause of disease in many cases. Another Hippocratic text (a passage from *The Art*) illustrates perfectly the difficulties experienced by learned physicians, who were regularly faced in practice with the inability to identify the cause of the disease, leading ultimately to failure in the management of the patient, particularly for *internal diseases*:

> More pains, in fact, and quite as much time, are required to know them as if they were seen with the eyes; for what escapes the eyesight is mastered by the eye of the mind, and the sufferings of patients due to their not being quickly observed are the fault, not of the medical attendants, but of the nature of the patient and of the disease. The attendant in fact, as he could neither see the trouble with his eyes nor learn it with his ears, tried to track it by reasoning. Indeed, even the attempted reports of their illnesses made to their attendants by sufferers from obscure diseases are the result of opinion, rather than of knowledge. If indeed they understood their diseases they would never have fallen into them, for the same intelligence is required to know the causes of diseases as to understand how to treat them with all the treatment that prevents illnesses from growing worse. Now when not even the reports afford perfectly reliable information, the attendant must look out for fresh light. For the delay thus caused not the art is to blame, but the constitution of human bodies. For it is only when the art sees its way that it thinks it right to give treatment, considering how it may give it, not by daring but by judgment, not by violence but by gentleness. As to our human constitution, if it admits of being seen, it will also admit of being healed. But if, while the sight is being won, the body is mastered by slowness in calling in the attendant or by the rapidity of the disease, the patient will pass away.[20]

This uneasiness, which was maintained by the repeated failures with which medicine was continually confronted, is also perceptible in other texts from the Hippocratic Corpus of the fifth century. It worsened in subsequent centuries, probably contributing to the emergence of several schools of medical thought, or "sects" as they were known, (when this was not a pejorative term) that defined themselves and opposed each other precisely on this question of the causes of diseases and the need for the doctor to know and to identify them. We know about these sects from the preface of *De Medicina* by Celsus and from *On the sects for beginners* by Galen, a major text that was read, reread, analysed, commented on and taught to young medical students until the eighteenth century. In this text, Galen discussed the methods used by the three sects to find remedies. He took a stand against the methodical and empirical sects, the latter he considered were only interested in the evident

---

[19] Hippocrates, Ancient Medicine 20 (translated by W. H. S. Jones, Cambridge, MA: Harvard University Press, The Loeb Classical Library Vol. 1, 1923, pp. 227–229).

[20] Hippocrates, The Art 11 (translated by W. H. S. Jones, Cambridge, MA: Harvard University Press, The Loeb Classical Library Vol. 2, 1923, pp. 209–211).

causes of disease and what he called the *theorem* of memory, repeating what had already worked several times before:

> But there is yet a third kind of experience, namely, the imitative one. An experience is imitative if something which has proved to be beneficial or harmful, either naturally or by chance or by extemporization, is tried out again for the same disease. It is this kind of experience which has contributed most to their art. For when they have imitated, not just twice or three times, but very many times, what has turned out to be beneficial on earlier occasions, and when they then find out that, for the most part, it has the same effect in the case of the same diseases, then they call such a memory a theorem and think that it already is trustworthy and forms part of the art. But when many such theorems had been accumulated by them, the whole accumulation amounted to the art of medicine, and the person who had accumulated the theorems, to a doctor.[21]

Galen himself favoured searching for causes, so that diseases could be treated by opposing the cause. In the same text, a little further on, he attacked the empirical sect directly, concerning the example of two patients bitten by a rabid dog:

> If you do say, as I also have heard from you in the beginning, that all is not manifest is useless and if you agree to follow what is obvious, then, perhaps, I can point out to you what it is that you are overlooking, reminding you of what is apparent. Two men, bitten by a mad dog, went to their familiar doctor, asking to be cured. In both cases, the wound was small, so that the skin was not even entirely torn, and one of them only treated the wound, not busying himself with anything else, and, after a few days, the part affected seemed to be fine. But the other, since he knew that the dog was mad, far from hastening to have the wound scar, did exactly the opposite and tried constantly to enlarge it, using strong and sharp drugs, till, after a considerable amount of time, he also forced the patient at this point to drink the medicines appropriate for madness, as he himself explained. And this is the end the whole matter took in both cases. The one who drank the medicines was saved and became healthy again. The other thought that he was not suffering anything, but all of a sudden came to fear water, went into spasms and died. Do you think that, in such cases, one inquires in vain into the antecedent cause and that the man died for any reason than the negligence of the doctor, who failed to ask at all about the cause and to apply the treatment observed in this case? To me it seems that he died for no other reason than this.[22]

This all too brief return to the origins of learned medicine demonstrates the long-standing and persistent nature of the tensions between the "deterministic" approaches used in medical practice, particularly in diagnosis. These tensions soon led to the emergence of alternative, so-called "empirical" approaches—based on "memory alone" according to Galen—but which could easily today be seen as pre- or crypto-probabilistic, and which have been maintained (not always in a marginal position) in medicine throughout its history, right up to the present day, when they are expressed in the only slightly more structured form of clinical prediction rules, especially when there are promoted by zealous supporters of Evidence-Based

---

[21] Galen, On the Sects for beginners 2 (translated by R. Walzer and M. Frede, Indianapolis: Hackett Publishing, 1985, p. 4).

[22] Ibid., pp. 13–14.

Medicine.[23] These tensions could be seen as *inherent to medicine*, which is often described to be both "an art and a science",[24] combining both practice and the search for an understanding of living beings and their diseases, the practitioners of which are confronted daily with both failures of efficacy and difficulties in finding satisfactory answers to the questions from patients and society concerning the cause of diseases. Medicine, both in the domain of diagnosis and in other domains of practice, will no doubt continue well into the future to try to reconcile the overriding desire to be effective and the desire to explain, and to find a balance between these two objectives when it is not possible to satisfy both.

References

Ball GV (1971) Two epidemics of gout. Bull Hist Med 45:401–408

Burnside E (2003) Bayesian networks computer-assisted diagnosis support in radiology. Acad Radiol 12:422–430

Duffin J (2006) Jodocus Lomnius's little golden book and the history of diagnostic semiology. J Hist Med Allied Sci 61:249–287

Ebell M (2001) Evidence-based diagnosis. A handbook of clinical prediction rules. Springer, New York, p 2

Garrod AB (1859) The nature and treatment of gout and rheumatic gout. Walton and Moberly, London

Ransohoff DF, Feinstein AR (1976) Editorial: is decision analysis useful in clinical medicine? Yale J Biol Med 49:165–168

Schwartz WB, Wolfe HJ, Pauker SG (1981) Pathology and probabilities: a new approach to interpreting and reporting biopsies. N Engl J Med 305:917–923

Stiell IG, McKnight RD, Greenberg GH, McDowell I, Nair RC, Wells GA, Johns C, Worthington JR (1994) Implementation of the Ottawa ankle rules. JAMA 271:827–832

Tversky A, Kahneman D (1974) Judgment under uncertainty: heuristics and biases. Science 185:1124–1131

---

[23] How can one not be struck by the similarity of the collection "of a large number of memory theorems" favoured by the ancient empirical sect and the recent work of Mark Ebell (2001), which also presents a collection of probabilistic rules developed for diverse clinical situations?

[24] For a discussion of medicine as "a science and an art", see the paper "Objectivity, Scientificity, and the Dualist Epistemology of Medicine" by Thomas Cunningham in this volume.

Risk Factor and Causality in Epidemiology

Élodie Giroux

Abstract The scientific and public health claim that smoking is a cause of lung cancer or cardiovascular diseases dates back to the mid-1960s. Nevertheless smoking is neither a necessary nor a sufficient condition for lung cancer. One of the main indicators for causality is that, at the population level, smoking highly increases the probability of having lung cancer. A probabilistic concept of causation was developed by some philosophers that could have given conceptual support to epidemiological causal analysis and inference. Yet, it appears that the agreement on the causal status of specific risk factors did not necessarily lead to the adoption of a probabilistic concept of causation by epidemiologists.

In this paper I propose a historical analysis of the emergence of the risk factor concept in epidemiology with the objective of highlighting how the question of causality arose. Causal inference in epidemiology has been structured by the famous Bradford Hill's criteria that were developed in the context of the 'smoking-lung cancer' controversy in a pragmatic objective and spirit. Even if there were not analysis of the implicit concept of causation presupposed by these criteria, I will show that there are several interpretations of causation behind these criteria which are more or less assumed by epidemiologists. All this leads us to the question of pluralism or monism with regard to the nature of causality in epidemiology and more generally in biomedicine.

É. Giroux (✉)
Institut de Recherches Philosophiques de Lyon, Université Jean-Moulin Lyon 3, Lyon, France
e-mail: elodie.giroux@univ-lyon3.fr

P. Huneman et al. (eds.), *Classification, Disease and Evidence*, History,
Philosophy and Theory of the Life Sciences 7, DOI 10.1007/978-94-017-8887-8_9,
© Springer Science+Business Media Dordrecht 2015

Cigarette smoking is a risk factor for many diseases; the association with lung cancer is particularly high. But this association is neither necessary nor sufficient. Indeed a person can develop this disease even if he has never smoked and vice-versa. In spite of this, it is now widely accepted that smoking is *a* cause of the lung cancer, even if not *the* cause. But what should we understand here by the term 'cause'? What is the causal status of the notion of 'risk factor'? Initially defined by epidemiologists as a variable statistically associated to the occurrence of a disease, the notion is now regarded as a useful and heuristic tool to cope with the multifactoriality of a phenomenon in numerous fields of study (Perreti-Watel 2004). But its true role in causal explanation needs some elucidation. Is it a vague substitute for the concept of cause? Or does it refer to a specific concept of causality, i.e. its probabilistic interpretation? The use of this notion is often ambivalent in epidemiology. It is unevenly used as a synonym for 'determinant' and 'cause' or, to the contrary, clearly distinguished from a 'causal factor' and then designated as a 'simple risk marker' (Last 1995).

The objective of this paper is an attempt to clarify the relation between the notions of 'risk factor' and 'cause'. In the first part I will show how the notion of risk factor was constructed by chronic disease epidemiology in the context of the development of population studies for aetiological research: cohort and case-control studies. This development was in part due to the introduction of new statistical techniques in inference. Within this context, the 'risk factor' notion – defined as a statistical correlation – could be seen as illustrating the positivist abandonment of the concept of cause. In the second part, I shall show how the controversy over the interpretation of the statistical association between cigarette smoking and lung cancer led to the formalisation of 'guidelines' for causal inference in epidemiology. The third and last part explores the status of these guidelines and criteria which are currently used as a toolbox for aetiological analysis in epidemiology. In conclusion I will try to answer the following question: can we use these criteria as efficient methodological tools for causal inference without having first to agree on a specific account of causation?

Risk Factors: Modeling Multifactoriality and Prediction

Identifying Risk Factors by Epidemiological Methods

The origins of the expression 'risk factor' in the insurance industry have already been established (e.g. Rothstein 2003). In the 1920s, some life insurance companies in the U.S.A. used this expression to characterize individual conditions such as obesity or hypertension. Within the discipline of epidemiology, the notion was more particularly used and diffused by one of the pioneering cohort studies on cardiovascular disease (CVD), the Framingham Heart Study (Aronowitz 1998; Giroux 2006, 2008). Indeed it is often claimed that this notion was introduced into the medical field by the paper that published the results of the 6 year follow-up of the population in 1961 (Last 1995)[1]: elevated serum cholesterol levels, hypertension, and the electrocardiographic pattern of left ventricular hypertrophy were designated as 'risk factors' for CVD because of their significant positive statistical association to the incidence rate of CVD.

The interest in variables statistically associated with the occurrence of a disease began earlier than the twentieth century. In her book *Les Causes de la mort, histoire naturelle des facteurs de risque*, Anne Fagot-Largeault has well shown the eighteenth and nineteenth century origins of aetiological research based on probabilistic reasoning. Nevertheless, it was only in the middle of the twentieth century that an epidemiological notion of risk factor emerged and acquired a central place in aetiological research; its increasingly widespread use in medicine can be clearly dated back to the 1970s (Skolbekken 1995). Indeed, in the middle of the twentieth century, this notion, or more generally, aetiological research in epidemiology, acquired its theoretical and methodological foundations in the context of the development of population studies (case-control and cohort studies) whose methods were designed with statistical tools of the emergent mathematical statistics.

At this time, in the more developed countries, what would later be called by Omran (1971) the 'demographic transition' was becoming more and more visible: chronic and multifactorial diseases, primarily cancer and CVD, were becoming the prevalent diseases. Little was known about their mechanisms and physiopathology. The absence of precise knowledge on their pathogenesis, their long and progressive development and their apparent irreversibility led some physicians to seek out screening tools and preventive techniques. In parallel, the comparison of the death rates of different countries and the statistical association shown by the data of studies carried out by the insurance industry brought to the fore aetiological hypotheses such as way of life and diet. Indeed, CVD mortality for which the American way of life and diet were thought responsible was much higher in the USA than in Japan. But it was difficult to draw any solid inference from these statistical associations. On the one hand, insurance data were collected from a portion of the insured population which could not be considered as representative of the American population as a whole. On the other hand, the comparison of death rates, which is an 'ecologic'[2] comparison, cannot warrant solid inference.

The development of the method of individual-level studies such as case-control and cohort studies allowed for better control of various forms of bias and confounding factors and then, a more rigorous base for statistical inferences on aetiological hypotheses. These studies consist in the organised observation of controlled and well-defined populations. The validity of the comparison depends on the methodological rigor of the study design and, more precisely the control of selection and information

---

[1] One of the first medical uses of the expression of 'risk factor' is usually referred to a publication of Framingham Heart Study (Kannel et al. 1961). On the importance of this study in the constitution and diffusion of the 'risk factor approach', see Aronowitz (1998), Oppenheimer (2006).

[2] The term 'ecologic' is used to refer to the population-level of analysis which relies on summary measures of health. Ecologic studies are to be distinguished from individual level studies.

bias in the constitution of the study population. In a case-control study, people with a disease are matched with people who do not have the disease ('control') and the cases' history of exposure or other characteristics, prior to the onset of the disease, are recorded and compared to the 'controls'. In a cohort study, which is often prospective, people who do not have the disease are followed and a comparison is made between the incidence rate[3] of those who are exposed to a suspected characteristic and the incidence rate of those who are not exposed.

In this kind of studies, a measure of the risk association between the suspected factor and the disease outcome is then possible: this is the 'relative risk', the ratio of the incidence rate of exposed group to the incidence rate of non-exposed group.[4] This measure delivers quantitative information on the strength of the association. The statistical technique of the test of significance is used to confirm that the association is unlikely to be due to chance. For example, after a 20-year follow-up in a prospective cohort study, the British epidemiologists Richard Doll and Richard Peto (1976) established that the relative risk of death due to CVD for men who are less than 65 years old is: 1 for the non-smokers, 1,7 for the smokers of 1–14 cigarettes (which means that smoking from 1–14 cigarettes a day multiplies the risk by 1,7); 2,2 for the smokers of 15–24 cigarettes; 2,6 for the smokers of 25 cigarettes and more. As shown by Anne Fagot-Largeault (1989), the quantification and the possibility of ranking the importance and relative weight of the various factors which are identified in these studies are the main differences with the tools and devices formerly used for aetiological research in epidemiology. Furthermore, mathematical models applied to analyze the data collected in these studies allow the development of a useful approach to the complex causation of some specific chronic diseases and the prediction of the probability of their occurrence in individuals.

The Multivariate Equation: Modeling Multifactoriality and Predicting the Risk

It was in the study of CVD that the relevance of the risk factors approach in modeling multifactoriality was the most obvious. Indeed, if in lung cancer one aetiological hypothesis (cigarette smoking) soon prevailed, it was different for CVD in which we have to deal with a large number of factors, each having a limited part in the risk: hypertension, hypercholesterolemia, age, sex, smoking, diabetes, and so on. Mathematical models appeared to be necessary for analyzing the numerous data of large cardiovascular cohort studies: indeed contingency tables and cross tabulation, the traditional method, quickly proved insufficient for comparison and quantification purposes (Giroux 2008). Multivariate models were then used, adapted to the analysis of epidemiological data (Cornfield 1962). Regression equation can model a phenomenon (dependant variable Y) by a function of several independent variables (X): $Y = \alpha + \beta 1 X 1 + \ldots + \beta n X n$. It is then

---

[3] The incidence rate is the proportion of subjects who develop a disease within a specified time period.
[4] In the case-control study it is not possible to obtain a direct measure of incidence rate. But an equivalent of the relative risk can be calculated: the odds ratio.

possible to take into account the combination of a large number of factors in the risk and their interaction, to control the potential confounding factors and to weigh the independent contribution of each individual factor.

Yet it should be noted that while models indeed have the advantage of summarizing the data and analyzing them easily, they also entail a kind of abstraction and a certain number of constraints. Such models are based on the isolation of factors, the independent part they play in the risk being investigated. It is only after this that the potential interactions with other risk factors are studied and considered. But this isolation leads to the neglect of some elements or contextual factors that play an important part in the emergence of these chronic diseases (see for example Diez-Roux 1998 or Giroux 2012).

These models, which came from mathematical statistics, were first used for aetiological analysis. But very soon they were also used for predicting risk. Especially in the USA, the clinical investigators of CVD epidemiology primarily aimed to identify potential patients or the 'coronarian profile' before the CVD became manifest. The possibility of predicting personal risk by the combination of several risk factors of the same individual was particularly attractive to the clinicians. Whether the factors are causal or not, they can be used to predict risk and identify people at risk. Indeed the predictive value of a factor is not always symmetrical to its aetiological part. For example, even if weight is an important aetiological factor in CVD, it is almost useless for predictive equation. This leads to underline the difference between prediction and aetiological analysis and the prevalence of the objective of risk prediction in the emergence and dissemination of the notion of 'risk factor'.

Thus, in the historical and epistemological study of its method of modeling risk factors, epidemiology appears to be intrinsically pragmatic. The risk-factor approach may be characterised by two main features also used by Perreti-Watel to explain what he calls the 'epidemiologic paradigm' (Perreti-Watel 2004)[5]: a primacy of prediction on aetiological analysis and an efficient modeling of multifactoriality. This model of multifactoriality and risk-factor epidemiology is also mainly focused on individual and biological attributes (Giroux 2012). This is particularly true for the first risk factors of CVD and cancer (cigarette smoking, hypertension, hypercholesterolemia, and so on): but what about causality in this paradigm? Is this pragmatism of clinicians and epidemiologists linked with a positivist substitution of the notion of cause for that of correlation, as defended by one of the main founders of inferential statistics, Karl Pearson (1892) and at the same time, by the philosopher Bertrand Russell (1912)?

Causal Inference: The Causal Status of Risk Factors

Despite the introduction of statistical techniques of inference and its increasing part in aetiological analysis in epidemiology and more widely in medicine, it does not seem that the causal language disappeared from those fields, to be replaced by the statistical

---

[5] In this paper, Perreti-Watel denounced the extension of this 'epidemiological paradigm' to the study of behaviors and individual beliefs.

notion of correlation. On the contrary, the identification of risk factors for disease quickly led to the question of their causal status. It seems that medicine needs some notion of cause.[6] Such persistence can be explained by the pragmatic orientation of aetiological research in medicine. For example, we are led to ask: which risk factors justify a preventive action? At what level of its aetiological part in the disease? When and how could a causal inference be justified? Thence the issue becomes that of causal inference and statistical methodology of proof: is it possible to prove causality from a statistical association? And if not, which level of evidence must we reach before deciding that the suspected causal link is high enough to justify a preventive action?

The 'smoking and lung cancer' controversy (1950–1964), mainly Anglo-American, played a central part in the analysis of these questions (Berlivet 2005). It is within this controversy that criteria and procedures for causal inference in epidemiology were established. Since the mid 1950s, several epidemiological studies (case-control studies and cohort studies) had shown the existence and strong statistical association between cigarette smoking and lung cancer. And yet there was no agreement among researchers as to whether this association was causal and justified a preventive action against smoking. The first difficulties to deal with were due to double evidence: first, there are non-smokers who have cancer and smokers who have not (neither necessity nor sufficiency) and second, smoking is statistically associated with many diseases other than lung cancer, such as cardiovascular and respiratory diseases (absence of specificity). Moreover no exact mechanism by which smoking causes cancer was known to make a causal relation plausible. It was therefore difficult to accept as causal a relation which was neither necessary nor sufficient, and for which we did not know any precise biological mechanism.

It should be noted that this controversy did not oppose statisticians to clinicians who were generally reluctant to adopt inferential statistics and statistical methods and reasoning: the statisticians themselves were divided. For them, the scientific recognition of the new statistical methodology and the virtue of its application in biology and medicine were also at stake: while promoting it, they nevertheless wanted to ensure its seriousness and to protect it from approximate medical use (Parascandola 2004). Two famous statisticians were sceptical and criticised the results of epidemiologic studies: Ronald A. Fisher and Joseph Berkson. For Berkson, the positive association observed in epidemiological studies could be attributed to a bias in the selection of individuals. For Fisher, a third factor (e.g. a genetic one) could predispose the individual both to smoke and to develop lung cancer: a confounding factor would thus be involved. To cast doubt on the claim of causality, Fisher relied on the very slight difference observed in the studies between the smokers who inhaled and those who did not. Such criticisms pushed to seek for an underlying biological mechanism that would better explain (or even, explain) the positive association between cigarette smoking and lung cancer.

The controversy led the public health services in England and in the USA to ask for reports from various experts. In 1962, the English report entitled 'Smoking and

---

[6] On the continuation and different forms of the notion of cause in the history of medicine, see Fagot-Largeault (1993). Concerning causal realism in medicine, see Grene (1976).

Health' concluded that there was a causal association. But it was more particularly the US Public Health Service's report published in 1964 that led to the standardisation of causal inference in epidemiology and ended the controversy.[7] This report also came to the conclusion that this risk factor had a causal status after a close debate on causal inference. A whole section was dedicated to the criteria of the epidemiologic judgment. It was stated that "statistical methods cannot establish proof of a causal relationship in an association", but that – coupled with data from clinical, experimental and pathological observations – the results of the epidemiologic studies provided a basis from which a causal judgment became possible. It was specified that this judgment went beyond any statistical probabilistic statement. There was no 'proof' but a judgment which relied on a set of criteria none of which was necessary or sufficient. The term 'cause' was said to be used here in its common meaning and not in its technical or philosophical sense. Five criteria[8] were used, to which the British epidemiologist Bradford Hill later added four (1965), thus making a list that rapidly became a kind of 'toolbox' for causal inference in epidemiology:

1. Strength of the risk association (the prospective studies showed that the incidence rates of lung cancer in the group of smokers were nine times higher than in the non-smokers)
2. Consistency (the experts of the Surgeon General noticed this correlation in 29 case-control studies and seven prospective cohort studies)
3. Specificity
4. Temporality
5. Biological Gradient
6. Plausibility
7. Coherence
8. Experiment
9. Analogy

The first two criteria tend to confirm the existence of the association and are based on statistical arguments. The following ones constitute various arguments of temporal, clinical and biological nature that help to determine if the association is causal. Epidemiologists thus use a cluster of heterogeneous clues to decide on the causal status of a risk factor. According to Anne Fagot-Largeault (1989) Hill's epidemiological criteria do no more than make explicit intuitive criteria; and this was stimulated by the context of the methodological refinement of study designs and aetiological analysis in epidemiology.[9]

---

[7] This report was the result of an analysis led by a committee of several experts appointed by the surgeon general. U.S. Department of Health Education and Welfare, Surgeon General's Report 1964. Smoking and Health: Report of the Advisory Committee to the Surgeon General of the Public Health Service. Washington DC: Government Printing Office.

[8] These five criteria were the following one: consistency of the association, strength of the association, specificity of the association, temporal relationship of the association, coherence of the association.

[9] According to Fagot-Largeault (1989), a causal explanation in medicine is a "judgment in which intervene in various proportion historical components (an aetiologic history), some calculus components (a statistical inference), and some decisional component (a choice relying on criteria of relevance)".

Bradford Hill had proposed these criteria as a pragmatic but non-dogmatic approach to causal inference. In his view it was not necessary that they all be met to decide on the causality of a relationship (Hill 1965). He insisted on the flexibility of their use and on the importance of adapting them to the context and matter of the study. But his aim was also to propose, beyond the controversy about smoking, a standardised procedure for causal inference in all sorts of disease. Today, his criteria are widely used in epidemiology and they are conventionally considered as the epidemiologic criteria of causality.

At this point, two questions arise about causal inference, and more generally about the status of causal judgment in epidemiology and medicine. The first one deals with the ability of this set of criteria to serve for any kind of disease. If they were useful and relevant for settling the 'smoking and health' controversy, is it also the case for CVD, the risk factors of which are not so strongly associated to their effect as cigarette smoking and lung cancer? Indeed the magnitude of the relative risk of most risk factors of this disease is lower. Controversies on the causal status of hypercholesterolemia were more difficult to settle, if at all. In this instance, socio-political, economic, industrial factors but also therapeutic findings are mere likely to intervene in the judgment of causality (Greene 2007).

Secondly, it might be thought that used as a toolbox of aetiological analysis these criteria in fact circumvent the more ontological issue concerning the nature and interpretation of the notion of cause. For Luc Berlivet (1995), they could be considered as a 'black box' that frees epidemiologists from the difficult question of defining the cause while answering the practical requirements of causal inference: 'the carving out of the concept of causality into criteria easy to understand offers the incomparable quality of being immediately working'. The contribution of Hill's criteria was then to have transferred the difficulty 'by turning a complicated and confusing debate into the formal procedure of checking one after the other the nine criteria. No philosophical debate any more, just classical laboratory process!' (Berlivet 1995, my translation). However can we really avoid the questions about the definition and interpretation of the concept of cause which underlie these operational criteria? This matter comes up in the debates and with it, the unity, or at least, consistency of the notion of causality in medicine.

Causal Interpretation

In the classical conception of scientific explanation in logical empiricism, the cause has been reduced to the premises of an explanation which has the form of a 'Deductive-Nomological' argument. There does not need to be any independent analysis of causality: the theory of explanation is sufficient to account for causality. Yet now, in epidemiology as in philosophy of science, there is a revival of the analysis of the notion of cause itself, in its ontological and conceptual dimensions. With regard to epidemiology, in an article entitled 'Causes', the American epidemiologist Kenneth Rothman underlined the importance of reducing the gap between the metaphysical conception of cause and epidemiological criteria of causal inference (1976). Mark Parascandola and Douglas Weed (2001), philosophers and epidemiologists at the National Institute of Cancer, assert that it is important to make explicit the criteria used but also the definitions of causality which are assumed by these criteria. For, though implicit, those definitions direct the way in which epidemiologists analyze the phenomena they study.

Does Risk Factor Imply a Probabilistic Concept of Cause?

To begin with, we must ask if the notion of risk factor implies a specific concept of cause. For Daniel Schwartz (1988), one of the main promoters of modern epidemiology in France, 'the statistical revolution leads to a new conception of cause: in the domain of probability, a cause is not expected to entail necessarily an effect but only to increase its probability'. At first sight, it seems that the use of statistical and probabilistic methods in epidemiology leads to the adoption of a probabilistic concept of cause. Such a concept would constitute an alternative to a causality conceived in logical terms of necessity and sufficiency. Various probabilistic theories of causation have been proposed and elaborated by philosophers including Hans Reichenbach (1956), Patrick Suppes (1970) and Ellery Eells (1991). They characterize the relationship between cause and effect using the tools of probability theory. The central idea is that causes raise the probabilities of their effects *ceteris paribus*.[10] Applied to epidemiological risk factors, this implies that a risk factor is considered causal because it increases the probability of its effect (the occurrence of such and such disease).

But it does not seem that this probabilistic conception of causation is the one assumed by the majority of epidemiologists. In spite of the explicit use made by some of such a notion of cause,[11] logic interpretation in terms of sufficiency and necessity and mechanistic account seems to prevail. Rothman (1976), for example, propounded a conception relying on the notions of necessity and sufficiency. His 'sufficient-component cause' definition is an adaptation of the notion of an INUS condition that had been proposed by the philosopher John Mackie (1965): an INUS condition for some effect is an Insufficient but Non-redundant part of an Unnecessary but Sufficient condition. While taking into account multifactoriality, Rothman thus maintains the notions of necessity and sufficiency in the causal analysis in epidemiology. A sufficient-component cause is indeed constituted by a set of components, none of which is in itself sufficient for the occurrence of the disease, but when all the components of a specific constellation or set are present, then the

---

[10] Most of the variation between probabilistic theory of causation and most of the debates are around the content of the *ceteris paribus* clause. The basic idea that causes raise the probability of their effects has indeed to be qualified to resolve the problem of spurious correlations and the problem of the symmetric nature of this simple 'probability-raising' condition.

[11] Elwood (1988), Lagiou et al. (2005), Parascandola and Weed (2001).

cause is sufficient. In Rothman's view, the probabilistic concept of cause could be useful – particularly in the context of public health and when there is emergency to decide with limited knowledge – but it is not yet precise enough and it does not give us any certain information about the individual. Nevertheless, as in the case of smoking and lung cancer, not all smokers develop lung cancer, and neither have we observed some specific set of conditions such that smoking is invariably followed by lung cancer in the presence of this set.

On the philosophers' side, criticisms of the probabilistic concept assert that probability theory is not enough to articulate a substantive account of causation. To Wesley Salmon (1984) for example, a continuous process, viewed as a real physical connection, must be identified. For him, only such a physical process allows us to distinguish among the statistical associations those which are causal.

Causal Concepts Underlying Epidemiological Causal Criteria: Pluralism, Reducibility or Complementarity?

In interpreting the notion of causation underlying the epidemiological criteria, we observe a tension between mechanistic considerations and statistical or probabilistic considerations; 'mechanistic' meaning what explains the occurrence of an effect and 'probabilistic' what makes a difference to the effect (Russo and Williamson 2007). The philosophers Federica Russo and Jon Williamson have analyzed the epidemiological criteria of causality and have showed that Hill's criteria 4, 6, 7, 8, 9[12] mentioned above involve mechanistic considerations while criteria 1, 2, 5, 8[13] involve probabilistic considerations. But can we be satisfied with such a dualism? This duality in types of causal considerations could be read as showing a causal pluralism in medicine. But it could also be read as a hidden supremacy of the mechanistic considerations, the probabilistic ones playing in fact just the heuristic role of pointing to potential causal hypotheses. In this latter view, it is argued that in the context of the 'smoking-lung cancer' controversy, in order to convince that the relation was causal, it was necessary to add the mechanistic considerations to the first two probabilistic criteria (strength and consistency) already fulfilled by the results of the epidemiological studies in the mid-1950s. These studies would have only played the role of indicating the most relevant causal hypothesis, the proof of causality being left to biological and pathological analyses which identify mechanisms.

But if we closely consider the debates on causal interpretation of risk factors, this rather reveals the relevancy of the thesis of an irreducible complementary relationship between mechanistic and probabilistic considerations. In the 'smoking-cancer' controversy, when the reports of 1962 then 1964 concluded a causal association, this was mainly based on the probabilistic type of considerations. This judgment

---

[12] (4) temporality, (6) plausibility, (7) coherence, (8) experiment, (9) analogy.

[13] (1) strength, (2) consistency, (5) biological gradient, (8) experiment.

was nevertheless also supported by a mechanistic consideration: a plausible mechanism of carcinogenesis was conceived. Some biological researchers then continued to seek a more direct cause, an agent which would be necessary and sufficient. The discovery in 1996 of the P53 gene revealed an even more accurate mechanism: the sequences of the gene which mutated in some cases of lung cancer are the same as those that tend to be related with the carcinogen molecule of tobacco named BPDE. But if this discovery strengthened the already accepted causal nature of the relation between smoking and lung cancer, it remained necessary to have information on the nature (positive or negative) and strength of the statistical association to ensure the *existence* of a causal relation.

The story of the 'cholesterol' hypothesis shows a similar to-and-fro between mechanistic and probabilistic considerations in the judgment of causality. Even if as early as 1913, thanks to the animal models and the experimental research led by Ludwig Aschoff (1866–1942) and Nicolai Anitschkov (1885–1964),[14] we had knowledge on plausible mechanisms of the role of blood cholesterol in the development of atherosclerosis, it was mainly the strength and the consistency of the statistical association between hypercholesterolemia and cardiovascular risk shown in epidemiologic studies that contributed to the conviction of the causal status of this risk factor. In 1994, the discovery of statins[15] allowed better understanding of the biological mechanisms at stake in this pathology. But the intervention epidemiological studies showing the efficiency of statins in terms of risk-benefit ratio were quite as important (Steinberg 2007). Thus, statistical evidence of epidemiological studies is indeed insufficient, but the knowledge of the biological mechanism is equally so. Several mechanisms can indeed link a factor to its potential effect. When a mechanism is already known, a causal inference is easier. But it remains necessary to study the statistical association and control the presence of other potential factors or confounding factors to determine the existence of the causal relation: other mechanisms could interfere in the same observed relation and the same mechanism could possibly lead to the same effect but from a different cause (Thagard 1998). Thus, statistical considerations determine the aetiology or the causal pathway through the identification of the relation from cause to effect. The mechanistic type of considerations identifies the pathogenesis or the intermediary process which explains the relation of cause to effect (Fagot-Largeault 1992).

The close complementary relation of these two aspects of causality in epidemiology leads us to consider the relevance of a unified account of causality in the health sciences. Several approaches have been proposed in this direction by philosophers of science. We shall briefly mention that of Paul Thagard (1999), followed by that of Russo and Williamson (2007). For Thagard, the unification concerns medical

---

[14] Anitschkov submitted rabbits to a diet mainly composed of eggs. The vasculary wall or intima of these rabbits were then covered with fat atherosclerotic layers thus considered as cholesterol (Anitschkov and Chalatow 1913).

[15] Statins are a kind of medicine which in acting on an enzyme of the metabolism pathway of cholesterol permit the decrease of the ratio of LDL-cholesterol (the 'bad cholesterol') in the blood.

explanation rather than medical theory of causation.[16] To him, the explanation in medicine is neither deductive nor statistical, nor in terms of single causes: it should be thought of as a 'causal network instantiation', 'where a causal network describes the interrelations among multiple factors, and instantiation consists of observational or hypothetical assignment of factors to the patient whose disease is being explained'. For each disease (cancer, ulcer, infectious disease, etc.), epidemiological studies and biological research establish a system of causal factors involved in the production of the disease. In this view, the unification of explanation is given by means of an organized collection of explanation schemas that characterize the causes of numerous diseases (Thagard 1999). The nodes of this causal network are connected by the causal relations inferred on the basis of several considerations: statistical associations, alternative causes, mechanisms (Thagard 1999). The making of the schema also relies on 'explanatory coherence'. Note that this notion of coherence, though not analyzed, had already been used by earlier epidemiologists (Susser 1973; Elwood 1988).[17]

On their side, Russo and Williamson defend a dual-faceted (or even multi-faceted) epistemic theory of causality 'as a unified account transcending the mechanistic and probabilistic accounts'. Both probabilistic and mechanistic aspects are crucial in health sciences when deciding whether or not to accept a causal assertion, but a single causal claim is used. Their objective is to account for the homogeneity of causal language in health sciences. For them, both monistic and pluralistic accounts of causality face epistemological problems. The dilemma is a false one due to a false dichotomy of monistic accounts into mechanistic and probabilistic, and due to the fact that these accounts confuse the types of evidence from which a causal assertion is drawn and the causal relation itself. There could exist two types of evidence for causal assertion and yet only one causal relation. In their view, the causal relation is not an ontological entity, it is epistemological: it should be identified with the causal beliefs of an omniscient rational agent. Causality is thus determined by causal epistemology. The duality or plurality in the types of evidence is thus compatible with a unified epistemic account of causality. In this approach of causality, there is a close link between conceptual analysis and epistemological analysis. The relevance and validity of this view remain to be examined. This research field is booming and sets epidemiology at the core of a reflection fundamental for philosophers of science.

To conclude, our historical and epistemological analysis of the relation between the notion of risk factor and that of cause has led us to highlight the important part that epidemiology plays in the development of criteria for causal inference but also in the renewal of reflections on the concept and theory of causation. On the boundaries between social sciences and biological sciences, epidemiology and its research

---

[16] Thagard explains that he does not seek here to define cause in terms of explanation or explanation in terms of cause. To him, causes, mechanisms, explanations, and explanatory coherence are intertwined notions.

[17] Thagard (1999) defined the 'explanatory coherence' as a positive constraint between hypotheses such as if one is accepted the other is too, and vice versa.

methods in disease causation present the opportunity for fruitful reflection on this matter which is so central for the philosophy of science. A convergence between the analysis of epidemiologists and statisticians and that of philosophers heralds a promising new era (See e.g. Pearl 2001; Rothman and Greenland 2005; Broadbent 2009).

References

Anitschkov N, Chalatow SS (1913) Über experimentelle Cholesterinsteatose und ihre Bedeutung für die Entstehung einiger pathologischer Prozesse. Zentralblatt für allgemeine Pathologie und pathologische Anatomie 24:1–9

Aronowitz R (1998) Making sense of illness: science, society and disease. Cambridge University Press, Cambridge

Berlivet L (1995) Controverse en épidémiologie. Production et circulation de statistiques médicales, Rapport pour la MIRE. CNRS, Rennes

Berlivet L (2005) 'Association or causation?' The debate on the scientific status of risk factor epidemiology, 1947–c.1965. Clio Med 25:39–74

Broadbent A (2009) Causation and models of disease in epidemiology. Stud Hist Philos Biol Biomed Sci 40:302–311

Cornfield J (1962) Joint dependence of risk of coronary heart disease on serum cholesterol and systolic blood pressure: a discriminant function analysis. Fed Proc 21:58–61

Diez-Roux A (1998) Bringing the context back into epidemiology. Am J Public Health 88:216–222

Doll R, Peto R (1976) Mortality in relation to smoking: 20 years' observations on male British doctors. Br Med J 6051:1525–1536

Eells E (1991) Probabilistic causality. Cambridge University Press, Cambridge

Elwood MJ (1988) Causal relationships in medicine. Oxford Medical Publication, Oxford

Fagot-Largeault A (1989) Les causes de la mort, histoire naturelle des facteurs de risque. Vrin, Paris

Fagot-Largeault A (1992) Quelques implications de la recherche étiologique. Sciences Sociales et Santé 10:33–45

Fagot-Largeault A (1993) On medicine's scientificity – did medicine's accession to scientific 'positivity' in the course of nineteenth century require giving up causal (etiological) explanation? In: Delkeskamp-Hayes C, Gardell Cutter MA (eds) Science, technology and the art of medicine. Kluwer Academic Publishers, Dordrecht, pp 105–126

Giroux E (2006) Épidémiologie des facteurs de risque: genèse d'une nouvelle approche de la maladie, thèse de doctorat en philosophie de la médecine. Université de Paris 1 Panthéon Sorbonne, Paris

Giroux E (2008) Enquête de cohorte et analyse multivariée: une analyse épistémologique et historique du rôle fondateur de l'étude de Framingham. Rev Epidemiol Sante Publique 56:177–188

Giroux E (2012) The Framingham Study and the constitution of a restrictive concept of risk factor. Soc Hist Med. doi:10.1093/shm/hks051

Greene JA (2007) Prescribing by numbers: drugs and the definition of disease. The Johns Hopkins University Press, Baltimore

Grene M (1976) Philosophy of medicine: prolegomena to a philosophy of science. In: PSA: Proceedings of the biennial meeting of the Philosophy of Science Association. The University of Chicago Press, pp 77–93

Hill BA (1965) Environment and disease: association or causation? Proc R Soc Med 58:295–300

Kannel WB, Dawber T, Kagan A, Revotskie N, Stokes JI (1961) Factors of risk in the development of coronary heart disease, six-year follow-up experience, the Framingham study. Ann Intern Med 55:33–48

Lagiou P, Adam HO, Trichopoulos D (2005) Causality, in cancer epidemiology. Eur J Epidemiol 20:565–574

Last JM (1995) A dictionary of epidemiology. Oxford University Press, New York/Oxford/Toronto

Mackie JL (1965) Causes and conditions. Am Philos Q 2:245–264

Omran AR (1971) The epidemiologic transition – a theory of the epidemiology of population change. Milbank Meml Fund Q 49(4):509–538

Oppenheimer GM (2006) Profiling risk: the emergence of coronary heart disease epidemiology in the United States (1947–70). Int J Epidemiol 35(3):720–730

Parascandola M (2004) Skepticism, statistical methods, and the cigarette: a historical analysis of a methodological debate. Perspect Biol Med 47:246–261

Parascandola M, Weed D (2001) Causation in epidemiology. Perspect Biol Med 55:905–912

Pearl J (2001) Causal inference in the health sciences: a conceptual introduction. Health Serv Outcomes Res Methodol 2:189–220

Pearson K (1892) The grammar of science. Walter Scott, London

Perreti-Watel P (2004) Du recours au paradigme épidémiologique pour l'étude des conduites à risque. Revue Française de Sociologie 45:103–132

Reichenbach H (1956) The direction of time. University of California Press, Berkeley/Los Angeles

Rothman KJ (1976) Causes. Am J Epidemiol 104:587–592

Rothman KJ, Greenland S (2005) Causation and causal inference in epidemiology. Am J Public Health 95:S144–S150

Rothstein W (2003) Public health and the risk factor, a history of an uneven medical revolution. University of Rochester Press, Rochester

Russell B (1912) On the notion of cause. Proc Aristot Soc 13:1–26

Russo F, Williamson J (2007) Interpreting causality in the health sciences. Int Stud Philos Sci 21:157–170

Salmon WC (1984) Scientific explanation and the causal structure of the world. Princeton University Press, Princeton

Schwartz D (1988) L'irrésolu. In: Lellouch J (ed) Présent et futur de l'épidémiologie. INSERM, Paris, pp 35–46

Skolbekken J-A (1995) The risk epidemic in medical journals. Soc Sci Med 40(3):291 305

Steinberg D (2007) The cholesterol wars: the Cholesterol skeptics versus the preponderance of evidence. Academic Press, New York

Suppes P (1970) A probabilistic theory of causality. North-Holland Publishing Company, Amsterdam

Susser M (1973) Causal thinking in the health sciences. Oxford University Press, New York

Thagard P (1998) Explaining disease: correlations, causes, and mechanisms. Minds Mach 8:61–78

Thagard P (1999) How scientists explain disease. Princeton University Press, Princeton

Herding QATs: Quality Assessment Tools for Evidence in Medicine

Jacob Stegenga

Abstract Medical scientists employ 'quality assessment tools' (QATs) to measure the quality of evidence from clinical studies, especially randomized controlled trials (RCTs). These tools are designed to take into account various methodological details of clinical studies, including randomization, blinding, and other features of studies deemed relevant to minimizing bias and error. There are now dozens available. The various QATs on offer differ widely from each other, and second-order empirical studies show that QATs have low inter-rater reliability and low inter-tool reliability. This is an instance of a more general problem I call the underdetermination of evidential significance. Disagreements about the strength of a particular piece of evidence can be due to different—but in principle equally good—weightings of the fine-grained methodological features which constitute QATs.

Introduction

The diversity of evidence in modern medicine is amazing. Many causal hypotheses in medicine, for instance, have evidence generated from experiments on cell and tissue cultures, experiments on laboratory animals (alive at first, then dead, dissected, and analyzed), results of mathematical models, data from epidemiological studies of human populations, data from controlled clinical trials, and meta-level summaries from systematic reviews based on techniques such as meta-analysis and social processes such as consensus conferences. Moreover, each of these kinds of evidence

J. Stegenga (✉)
Institute for the History and Philosophy of Science and Technology,
University of Toronto, 91 Charles Street West, Toronto M5S 1 K7, ON, Canada

Department of Philosophy, University of Utah,
215 South Central Campus Drive, Carolyn Tanner Irish Humanities Building,
4th Floor, Salt Lake City, UT 84112, USA
e-mail: jacob.stegenga@utoronto.ca; http://individual.utoronto.ca/jstegenga

P. Huneman et al. (eds.), *Classification, Disease and Evidence*, History,
Philosophy and Theory of the Life Sciences 7, DOI 10.1007/978-94-017-8887-8_10,
© Springer Science+Business Media Dordrecht 2015

has many variations. Epidemiological studies on humans, for instance, include case-control studies, retrospective cohort studies, and prospective cohort studies.

Evidence from each of these diverse kinds of methods has varying degrees of credibility and relevance for a hypothesis of interest. It is crucial, in order to determine how compelling the available kinds of evidence are, and to make a well-informed assessment of a causal hypothesis, that one take into account substantive details of the methods that generated the available evidence for that hypothesis. Methodological quality, in medical research at least, is typically defined as the extent to which the design, conduct, analysis, and report of a medical trial minimizes potential bias and error. Medical scientists attempt to account for the various dimensions of quality of evidence in a number of ways.

Methodological quality is a complex multi-dimensional property that one cannot simply intuit, and so formalized tools have been developed to aid in the assessment of the quality of medical evidence. Medical evidence is often assessed rather crudely by rank-ordering the types of methods according to an 'evidence hierarchy'. Systematic reviews and specifically meta-analyses are typically at the top of such hierarchies, randomized controlled trials are near the top, non-randomized cohort and case-control studies are lower, and near the bottom are laboratory studies and anecdotal case reports.[1] Evidence from methods at the top of this hierarchy, especially evidence from clinical trials, is often assessed with more fine-grained tools that I call 'quality assessment tools' (QATs). There are many such tools now on offer—QATs are quickly becoming an important tool of medical scientists. QATs are used to assess the primary-level evidence amalgamated by a systematic review, and since most causal hypotheses in medicine are assessed by evidence generated from systematic reviews, much of what we think we know about causal hypotheses in medicine is influenced by QATs.

A widely accepted norm holds that when determining the plausibility of a hypothesis one should take into account all of the available evidence. For hypotheses about medical interventions this principle stipulates that one ought to take into account the range of diverse kinds of evidence which are available for that hypothesis.[2] A similar norm states that when determining the plausibility of a hypothesis one should take into account how compelling the various kinds of evidence available for that hypothesis are, by considering detailed qualitative features of the methods used

---

[1] I discuss evidence hierarchies in more detail below. Such evidence hierarchies are commonly employed in evidence-based medicine. Examples include those of the Oxford Centre for Evidence-Based Medicine, the Scottish Intercollegiate Guidelines Network (SIGN), and The Grading of Recommendations Assessment, Development and Evaluation (GRADE) Working Group. These evidence hierarchies have recently received much criticism. See, for example, Bluhm (2005), Upshur (2005), Borgerson (2008), and La Caze (2011), and for a specific critique of placing meta-analysis at the top of such hierarchies, see Stegenga (2011). In footnote 5 below I cite several recent criticisms of the assumption that RCTs ought to be necessarily near the top of such hierarchies.

[2] The general norm is usually called the principle of total evidence, associated with Carnap (1947). See also Good (1967). Howick (2011) invokes the principle of total evidence for systematic reviews of evidence related to medical hypotheses. A presently unpublished paper by Bert Leuridan contains a good discussion of the principle of total evidence as it applies to medicine.

to generate that evidence. The purpose of using a QAT is to evaluate quality of evidence in such a fine-grained way.

A burgeoning literature has investigated the strategies that scientists employ when generating and assessing evidence.[3] In what follows I examine the use of QATs as codified tools for assessing evidence in medical research. Although there has been some criticism of QATs in the medical literature, they have received little philosophical critique.[4] I begin by describing general properties of QATs, including the methodological features that many QATs share and how QATs are typically employed. I then turn to a discussion of empirical studies which test the inter-rater reliability and inter-tool reliability of QATs. Although I refrain from evaluation of any particular QAT, I defend their general use in medical research. However, most QATs are not very good at constraining intersubjective assessments of hypotheses, and more worrying, the use of different QATs to assess the same primary evidence leads to widely divergent quality assessments of that evidence, which is an instance of a more general problem I call the underdetermination of evidential significance. This thesis holds that in a rich enough empirical situation, the strength of the evidence is underdetermined.

Quality Assessment Tools

A quality assessment tool (QAT) for medical evidence can be either a scale with elements that receive a quantitative score representing the degree to which each element is satisfied by a medical trial, or else a QAT can be simply a checklist with elements that are marked as either present or absent in a medical trial. Given the emphasis on randomized controlled trials (RCTs) in medical research, most QATs are designed for the evaluation of RCTs, although there are several for observational studies and systematic reviews.[5] Most QATs share several elements, including questions about how subjects were assigned to experimental groups in a trial, whether or not the subjects and experimenters were 'blinded' to the subjects' treatment protocol, whether or not there was a sufficient description of subject withdrawal from the trial

---

[3] See, for example, Hacking (1981), Thagard (1998), Bechtel (2002), and Weber (2009).

[4] Although one only needs to consider the prominence of randomization in QATs to see that QATs have, in fact, been indirectly criticized by the recent literature criticizing the assumed 'gold standard' status of RCTs (see footnote 5). In the present paper I do not attempt a thorough normative evaluation of any particular QAT. Considering the role of randomization suggests what a large task a thorough normative evaluation of a particular QAT would be. But for a systematic survey of the most prominent QATs, see West et al. (2002).

[5] The view that RCTs are the 'gold standard' of evidence has recently been subjected to much philosophical criticism. See, for example, Worrall (2002, 2007), Cartwright (2007), and Cartwright (2010); for an assessment of the arguments for and against the gold standard status of RCTs, see Howick (2011). Observational studies also have QATs, such as QATSO (Quality Assessment Checklist for Observational Studies) and NOQAT (Newcastle-Ottawa Quality Assessment Scale – Case Control Studies).

groups, whether or not particular statistical analyses were performed, and whether or not a report of a trial disclosed financial relationships between investigators and companies.[6] Most QATs provide instructions on how to score the individual components of the QAT and how to determine an overall quality score of a trial.

A comprehensive list of QATs developed by the mid-1990s was described by Moher et al. (1995). The first scale type to be developed, known as the Chalmers scale, was published in 1981. By the mid-1990s there were over two dozen QATs, and by 2002 West et al. were able to identify 68 for RCTs or observational studies. Some are designed for the evaluation of any medical trial, while others are designed to assess specific trials, or trials from a particular medical sub-discipline. Some are designed to assess the quality of a trial itself, while others are designed to assess the quality of a report of a trial, but most assess both.

QATs are now widely used for several purposes. When performing a systematic review of the available evidence for a particular hypothesis, QATs help reviewers take the quality of medical studies into account. This is typically done in one of two ways. First, QAT scores can be used to generate a weighting factor for the technique known as meta-analysis. Meta-analysis usually involves calculating a weighted average of so-called effect sizes from individual medical studies, and the weighting of effect sizes can be determined by the score of the respective trial on a QAT.[7] Second, QAT scores can be used as an inclusion criterion for a systematic review, in which any primary-level trial that achieves a QAT score above a certain threshold would be included in the systematic review (and conversely, any trial that achieves a QAT score below such a threshold would be excluded). This application of QATs is perhaps the most common use to which they are put. Finally, QATs can be used for purposes not directly associated with a particular systematic review or meta-analysis, but rather to investigate relationships between QAT scores and other properties of medical trials. For instance, several findings suggest that there is an inverse correlation between QAT score and effect size (in other words, higher quality trials tend to have lower estimates of the efficacy of medical interventions).[8]

Why should medical scientists bother using QATs to assess evidence? Consider the following argument, similar to an argument for following the principle of total evidence, based on a concern to take into account any 'defeating' properties of one's evidence. Suppose your evidence seems to provide definitive support for some hypothesis, H_1. But then you learn that there is a systematic error in the method

---

[6] A note about terminology: sometimes the term 'trial' in the medical literature refers specifically to an experimental design (such as a randomized controlled trial) while the term 'study' refers to an observational design (such as a case control study), but this use is inconsistent. I will use both terms freely to refer to any method of generating evidence in biomedical research, including both experimental and observational designs.

[7] There are several commonly employed measures of effect size, including mean difference (for continuous variables), or odds ratio, risk ratio, or risk difference (for dichotomous variables). The weighting factor is sometimes determined by the QAT score, but a common method of determining the weight of a trial is simply based on the size of the trial (Egger et al. 1997), often by using the inverse variability of the data from a trial to measure that trial's weight (because inverse variability is correlated with trial size).

[8] See, for example, Moher et al. (1998), Balk et al. (2002), and Hempel et al. (2011).

which generated your evidence. Taking into account this systematic error, the evidence no longer supports H_1 (perhaps instead the evidence supports a competitor hypothesis, H_2). Had you not taken into account the fine-grained methodological information regarding the systematic error, you would have unwarranted belief in H_1. You don't want to have unwarranted belief in a hypothesis, so you'd better take into account fine-grained methodological information.[9]

Here is a related argument: if one does not take into account all of one's evidence, including one's old evidence, then one is liable to commit the base-rate fallacy. In terms of Bayes' Theorem—$p(H|e) = p(e|H)p(H)/p(e)$—one commits the base-rate fallacy if one attempts to determine $p(H|e)$ without taking into account $p(H)$. Similarly, if one wants to determine $p(H|e)$ then one ought to take into account the detailed methodological features which determine $p(e|H)$ and $p(e)$.

One need not be a Bayesian to see the importance of assessing evidence at a fine-grain with QATs. For instance, Mayo's notion of 'severe testing', broadly based on aspects of frequentist statistics, also requires taking into account fine-grained methodological details. The Severity Principle, to use Mayo's term, claims that "passing a test T (with e) counts as a good test of or good evidence for H just to the extent that H fits e and T is a *severe test* of H" (Mayo 1996).[10] Attending to fine-grained methodological details to ensure that one has minimized the probability of committing an error is central to ensuring that the test in question is severe, and thus that the Severity Principle is satisfied. So, regardless of one's doctrinal commitment to Bayesianism or frequentism, the employment of tools like QATs to take into account detailed information about the methods used to generate the available evidence ought to seem reasonable.

One of the simplest QATs is the Jadad scale, first developed in the 1990s to assess clinical studies in pain research. Here it is, in full:

1. Was the study described as randomized?
2. Was the study described as double blind?
3. Was there a description of withdrawals and dropouts?

A 'yes' to question 1 and question 2 is given one point each. A 'yes' to question 3, in addition to a description of the number of withdrawals and dropouts in each of the trial sub-groups, and an explanation for the withdrawals or dropouts, receives one point. An additional point is given if the method of randomization is described in the paper, and the method is deemed appropriate. A final point is awarded if the method of blinding is described, and the method is deemed appropriate. Thus, a trial can receive between zero and five points on the Jadad scale.

---

[9] The parallel argument for the principle of total evidence is based on a concern to avoid 'defeating' evidence. Defeating evidence has the following property. Suppose some hypothesis H is confirmed by some piece of evidence (e_c). Then some other piece of evidence (e_d) is defeating if $p(H|e_c \& e_d) < p(H|e_c)$. This could arise, for instance, because e_d provides strong reason to believe that e_c is, in fact, spurious.

[10] The latter notion—H passing a severe test T with x_0—occurs when "1) x_0 agrees with H, (for a suitable notion of 'agreement') and 2) with very high probability, test T would have produced a result that accords less well with H than does x_0, if H were false or incorrect" (Mayo and Spanos 2011).

Table 1 Number of methodological features used in six QATs, and weight assigned to three widely shared methodological features

Scale	Number of items	Weight of randomization	Weight of blinding	Weight of withdrawal
Chalmers et al. (1981)	30	13.0	26.0	7.0
Jadad et al. (1996)	3	40.0	40.0	20.0
Cho and Bero (1994)	24	14.3	8.2	8.2
Reisch et al. (1989)	34	5.9	5.9	2.9
Spitzer et al. (1990)	32	3.1	3.1	9.4
Linde et al. (1997)	7	28.6	28.6	28.6

Adapted from Jüni et al. (1999)

The Jadad scale has been praised by some as being easy to use—it takes about 10 min to complete for each study—which is an obvious virtue when a reviewer must assess hundreds of studies for a particular hypothesis. On the other hand, others complain that it is too simple, and that it has low inter-rater reliability (discussed in section "Inter-rater reliability"). I describe the tool here not to assess it but merely to provide an example of a QAT for illustration.

In contrast to the simplicity of the Jadad scale, the Chalmers scale has 30 questions in several categories, which include the trial protocol, the statistical analysis, and the presentation of results. Similarly, the QAT developed by Cho and Bero (1994) has 24 questions. At a coarse grain some of the features on the Chalmers QAT and the Cho and Bero QAT are similar to the basic elements of the Jadad QAT: these scales both include questions about randomization, blinding, and subject withdrawal. (In section "Underdetermination of evidential significance" I briefly describe how Cho and Bero developed their QAT, as an illustration of the no-best-weighting argument). In addition, these more detailed QATs include questions about statistical analyses, control subjects, and other methodological features deemed relevant to minimizing systematic error. These QATs usually take around 30–40 min to complete for each study. Despite the added complexity of these more detailed QATs, their scoring systems are kept as simple as possible. For instance, most of the questions on the Cho and Bero QAT allow only the following answers: 'yes' (2 points), 'partial' (1 point), 'no' (0 points), and 'not applicable' (0 points). This is meant to constrain the amount of subjective judgment required when generating a QAT score.

Although most QATs share at least several similar features, the relative weight of the overall score given to the various features differs widely between QATs. Table 1 lists the relative weight of three central methodological features—subject randomization, subject allocation concealment (or 'blinding'), and description of subject withdrawal—for the above QATs, in addition to three other QATs.

Note two aspects of Table 1. First, the number of items on a QAT is highly variable, from 3 to 34. Second, the weight given to particular methodological features is also highly variable. Randomization, for instance, constitutes 3.1 % of the overall score on the QAT designed by Spitzer et al. (1990), whereas it constitutes 40 % of the overall score on the QAT designed by Jadad et al. (1996). The differences between

QATs explains the low inter-tool reliability, which I describe in section "Inter-tool reliability". But first I describe the low inter-rater reliability of QATs.

Inter-rater Reliability

The extent to which multiple users of the same rating system achieve similar ratings is usually referred to as 'inter-rater reliability'. Empirical evaluations of the inter-rater reliability of QATs have shown a wide disparity in the outcomes of a QAT when applied to the same primary-level study by multiple reviewers; that is, the inter-rater reliability of QATs is, usually, poor.

The typical set-up of evaluations of inter-rater reliability of a QAT is simple: give a set of manuscripts to multiple reviewers who have been trained to use the QAT, and compare the quality scores assigned by these reviewers to each other. A statistic called kappa (κ) is typically computed which provides a measure of agreement between the quality scores produced by the QAT from the multiple reviewers (although other statistics measuring agreement are also used, such as Kendall's coefficient of concordance and the intraclass correlation coefficient).[11] Sometimes the manuscripts are blinded as to who the authors were and what journals the manuscripts were published in, but sometimes the manuscripts are not blinded, and sometimes both blinded and non-blinded manuscripts are assessed to evaluate the effect of blinding. In some cases the manuscripts all pertain to the same hypothesis, while in other cases the manuscripts pertain to various subjects within a particular medical sub-discipline.

For example, Clark et al. (1999) assessed the inter-rater reliability of the Jadad scale, using four reviewers to evaluate the quality of 76 manuscripts of RCTs. Inter-rater reliability was found to be "poor", but it increased substantially when the third item of the scale (explanation of withdrawal from study) was removed and only the remaining two questions were employed.

A QAT known as the 'risk of bias tool' was devised by the Cochrane Collaboration (a prominent organization in the so-called evidence-based medicine movement) to assess the degree to which the results of a study "should be believed." A group of medical scientists subsequently assessed the inter-rater reliability of the risk of bias tool. They distributed 163 manuscripts of RCTs among five reviewers, who assessed the RCTs with this tool, and they found the inter-rater reliability of the quality assessments to be very low (Hartling et al. 2009).

---

[11] For simplicity I will describe Cohen's Kappa, which measures the agreement of two reviewers who classify items into discrete categories, and is computed as follows:

$$\kappa = [p(a) - p(e)]/[1 - p(e)]$$

where p(a) is the probability of agreement (based on the observed frequency of agreement) and p(e) is the probability of chance agreement (also calculated using observed frequency data). Kappa was first introduced as a statistical measure by Cohen (1960). For more than two reviewers, a measure called Fleiss' Kappa can be used. I give an example of a calculation of κ below.

Similarly, Hartling et al. (2011) used three QATs (Risk of Bias tool, Jadad scale, Schulz allocation concealment) to assess 107 studies on a medical intervention (the use of inhaled corticosteroids for adults with persistent asthma). This group employed two independent reviewers who scored the 107 studies using the three QATs. They found that inter-rater reliability was 'moderate'. However, the claim that inter-rater reliability was moderate was based on a standard scale in which a κ measure between 0.41 and 0.6 is deemed moderate. The κ measure in this paper was 0.41, so it was just barely within the range deemed moderate. The next lower category, with a κ measure between 0.21 and 0.4, is deemed 'fair' by this standard scale. But at least in the context of measuring inter-rater reliability of QATs, a κ of 0.4 represents wide disagreement between reviewers.

Here is a toy example to illustrate the disagreement that a κ measure of 0.4 represents. Suppose two teaching assistants, Beth and Sara, are grading the same class of 100 students, and must decide whether or not each student passes or fails. Their joint distribution of grades is:

		Sara	
		Pass	Fail
Beth	Pass	40	10
	Fail	20	30

Of the 100 students, they agree on passing 40 students and failing 30 others, thus their frequency of agreement is 0.7. But the probability of random agreement is 0.5, because Beth passes 50 % of the students and Sara passes 60 % of the students, so the probability that Beth and Sara would agree on passing a randomly chosen student is 0.5×0.6 (= 0.3), and similarly the probability that Beth and Sara would agree on failing a randomly chosen student is 0.5×0.4 (= 0.2) (and so the overall probability of agreeing on passing or failing a randomly chosen student is $0.3 + 0.2 = 0.5$). Applying the kappa formula gives:

$$(0.7 - 0.5) / (1 - 0.5) = 0.4$$

Importantly, Beth and Sara disagree about 30 students regarding a relatively simple property (passing). It is natural to suppose that they disagree most about 'borderline' students, and their disagreement is made stark because Beth and Sara have a blunt evaluative tool (pass/fail grades rather than, say, letter grades). But a finer-grained evaluative tool would not necessarily mitigate such disagreement, since there would be more categories about which they could disagree for each student; a finer-grained evaluative tool would increase, rather than decrease, the number of borderline cases (because there are borderline cases between each letter grade). This example is meant to illustrate that a κ measure of 0.4 represents poor agreement between two reviewers.[12] A κ score is fundamentally an arbitrary measure of disagreement, and the significance of the disagreement that a particular κ score represents presumably varies with context. This example, I nevertheless hope, helps to illustrate the extent of disagreement found in empirical assessments of the inter-rater reliability of QATs.

In short, different users of the same QAT, when assessing the same evidence, generate diverging assessments of the strength of that evidence. In most tests of the inter-rater reliability of QATs, the evidence being assessed comes from a narrow range of study designs (usually all the studies are RCTs), and the evidence is about a narrow range of subject matter (usually all the studies are about the same causal hypothesis regarding a particular medical intervention). The poor inter-rater reliability is even more striking considering the narrow range of study designs and subject matter from which the evidence is generated.

Inter-tool Reliability

The extent to which multiple instruments have correlated measurements when applied to the same property being measured is referred to as inter-tool reliability. A QAT has inter-tool reliability with respect to another QAT if its measurement of the quality of medical studies correlates with the measurement of the quality of the same studies by the other QAT. Because the score from a QAT is measured on a relatively arbitrary scale, and because the scales between multiple QATs are incommensurable, constructs such as 'high quality' and 'low quality' are developed for each QAT which allow the results from different QATs to be compared. That is, when testing the inter-tool reliability of multiple QATs, what is usually being compared is the extent of agreement among the QATs regarding the categorization of particular medical trials into pre-defined bins of quality. Similar to assessments of inter-rater reliability, empirical evaluations of the inter-tool reliability have shown a wide disparity in the outcomes of multiple QATs when applied to the same primary-level studies; that is, the inter-tool reliability of QATs is poor. I should note, however, that there are few such assessments of the inter-tool reliability of QATs, and

---

[12] I owe Jonah Schupbach thanks for noting that a κ measure can not only seem inappropriately low, as in the above cases of poor inter-rater reliability, but can seem inappropriately high as well. If a κ measure approaches 1, this might suggest agreement which is 'too good to be true'. Returning to my toy example, if Beth and Sara had a very high a κ measure, then one might wonder if they colluded in their grading. Thus when using a κ statistic to assess inter-rater reliability, we should hope for a κ measure above some minimal threshold (below which indicates too much disagreement) but below some maximum threshold (above which indicates too much agreement). What exactly these thresholds should be are beyond the scope of this paper (and are, I suppose, context sensitive).

those published thus far have varied with respect to the particular QATs assessed, the design of the reliability assessment, and the statistical analyses employed.[13]

An extensive investigation of inter-tool reliability was performed by Jüni and colleagues (1999). They amalgamated data from 17 studies which had tested a particular medical intervention (the use of low molecular weight heparin to prevent post-operative thrombosis), and they used 25 QATs to assess the quality of these 17 studies (thereby effectively performing 25 meta-analyses). The QATs that this group used were the same that Moher et al. (1995) had earlier described, which varied in the number of assessed study attributes, from a low of three attributes to a high of 34, and varied in the weight given to the various study attributes. Jüni and his colleagues noted that "most of these scoring systems lack a focused theoretical basis." Their results were troubling: the amalgamated effect sizes between these 25 meta-analyses differed by up to 117 %—*using exactly the same primary evidence*. They found that medical trials deemed high quality according to one QAT could be deemed low quality according to another QAT. The authors concluded that "the type of scale used to assess trial quality can dramatically influence the interpretation of meta-analytic studies."

Perhaps the most recent evaluation of inter-tool reliability is Hartling et al. (2011), discussed above in section "Inter-rater reliability". Recall that this group used three QATs (Risk of Bias tool, Jadad scale, Schulz allocation concealment) to assess 107 trials on a particular medical intervention. They also found that the inter-tool reliability of these QATs was very low.

Yet another example of a test of inter-tool reliability of QATs was reported by Moher et al. (1996). This group used six QATs to evaluate 12 trials of a medical intervention. Again, the inter-tool reliability was found to be low.

Low inter-tool reliability of QATs is troubling: it is a quantitative empirical demonstration that the determination of the quality of a medical trial depends on the choice of QAT. Moreover, in section "Quality assessment tools" I noted that there are many QATs with large differences between them. Thus the *best* methods that medical scientists have to determine the strength of evidence generated by what are typically deemed the best-designed medical studies (RCTs) are relatively unconstraining and liable to produce conflicting assessments.

Such low inter-tool reliability might be less troubling if the various QATs had distinct domains of proper application. The many biases present in medical research are pertinent to varying degrees depending on the details of the particular circumstances at hand, and so one might think it a mistake to expect that one QAT ought to apply to all circumstances. For some causal hypotheses, for instance, it is difficult or impossible to conceal the treatment from the subject (that is, 'blinding'

---

[13] For this latter reason I refrain from describing or illustrating the particular statistical analyses employed in tests of the inter-tool reliability of QATs, as I did in section "Inter-rater reliability" on tests of the inter-rater reliability of QATs. Nearly every published test of inter-rater reliability uses a different statistic to measure agreement of quality assessment between tools. Analyses employed include Kendall's rank correlation coefficient (τ), Kendall's coefficient of concordance (W), and Spearman's rank correlation coefficient (ρ).

is sometimes impossible)—hypotheses regarding chiropractic spinal manipulation are a case in point. Thus no study relevant to such a hypothesis will score well on a QAT that gives a large weight to blinding. Such a QAT would be less sensitive to the presence or absence of sources of bias other than lack of blinding, relative to QATs that give little or no weight to blinding. In such a case one might argue that since the absence of blinding is fixed among the relevant studies, an appropriate QAT to use in this case should not give any weight to blinding, and would only ask about the presence of those properties of a study that might vary among the relevant studies. On the other hand, one might argue that since we have principled reasons for thinking that the absence of blinding can bias the results of a study, even among those studies that cannot possibly be blinded, an appropriate QAT to use in this case *should* evaluate the presence of blinding (in which case all of the relevant studies would simply receive a zero score on blinding), just as a QAT ought to evaluate the presence of blinding in a scenario in which the studies in fact can be blinded. The former consideration is an appeal to determining the *relative* quality between studies, and the latter consideration is an appeal to determining the *absolute* quality of studies. The latter consideration should be more compelling in most cases, since the typical use of QATs (as discussed above) is to help estimate the true efficacy of a medical intervention, and such estimates ought to take into account the full extent of the potential for biases, regardless of whether or not it was possible for the relevant studies to avoid such biases.

There are scenarios, though, in which we might have reasons to think that a property of a study that causes bias in other scenarios does not cause bias (or perhaps causes less bias) in these scenarios. For example, the placebo effect might be stronger in studies that are designed to assess the benefits of pharmaceuticals compared with studies that are designed to assess the harms of pharmaceuticals. Such a difference could be independently and empirically tested. If this were so, then the different scenarios would indeed warrant different QATs, suitable for the particularities of the scenario at hand. If the low inter-tool reliability of QATs were merely the result of employing multiple QATs to different kinds of empirical scenarios (different kinds of studies, say, or studies of different kinds of hypotheses, such as benefits versus harms of pharmaceuticals), then such low inter-tool reliability would hardly be troubling. Indiscriminate use of QATs might lead to low inter-tool reliability of QATs, such thinking would go, but discriminate use of QATs will not.

Similarly, low inter-tool reliability of QATs would be less troubling if one could show that in principle there is only one good QAT for a given domain, or at least a small set of good QATs which are similar to each other in important respects, because then one could dismiss the observed low inter-tool reliability as an artefact caused by the inclusion of poor QATs in addition to good QATs.

Unfortunately, on the whole, these considerations do not mitigate the problem of low inter-tool reliability of QATs. There are, in fact, a plurality of equally fine QATs, designed for the same kinds of scenarios (typically: assessing RCTs of the efficacy of pharmaceuticals). A systematic review by medical scientists concluded that there were numerous QATs that "represent acceptable approaches that could be used today without major modifications" (West et al. 2002). Moreover, all of the empirical demonstrations of low inter-tool reliability of QATs involve the assessment of the quality of studies from a very narrow domain: for instance, the low inter-tool reliability of QATs shown in Jüni et al. (1999) involved assessing studies of a *single* design (RCTs) about a *single* causal hypothesis, and these QATs had been developed with the purpose of assessing the quality of that very study design. Although there are some QATs which are arguably inferior to others, at least among the reasonably good ones I argue below that we lack a theoretical basis for distinguishing among them, and so we are stuck with a panoply of acceptable QATs which disagree widely about the quality of particular medical studies and thus the strength of the evidence generated from those studies.

One might agree with the view that there is no uniquely best QAT, but be tempted to think that this is due only to the fact that the quality of a study depends on particularities of the context (e.g. the particular kind of study in question and the form of the hypothesis being tested by that study). Different QATs might, according to this thought, be optimally suited to different contexts. While this latter point is no doubt true—above I noted that some QATs are designed for assessing particular kinds of studies, and other QATs are designed for assessing studies in a particular domain of medicine—it does not explain the low inter-tool reliability of QATs as demonstrated by the empirical results cited above. That is because, as above, the low inter-tool reliability of QATs is demonstrated in very narrow particular contexts. Moreover, the research groups that design QATs usually claim (explicitly) that their QATs are meant to be applicable to a given study design (usually RCTs) in almost any domain of medical research. In short, QATs are intended to apply to a broad range of contexts, but regardless, the empirical demonstrations of their low inter-tool reliability are almost always within a single particular context.

Despite their widespread and growing use, among medical scientists there is some debate about whether or not QATs ought to be employed at all (see, for example, Herbison et al. (2006)). Their low inter-rater and inter-tool reliability might suggest that resistance to their use is warranted. There are three reasons, however, that justify the continuing improvement and application of QATs to assessing the quality of medical evidence. First, when performing a meta-analysis, a decision to not use an instrument to differentially weight the quality of the primary-level studies is equivalent to weighting all the primary-level studies to an equal degree. So whether one wishes to or not, when performing a meta-analysis one is forced, in principle, to weight the primary-level studies, and the remaining question then is simply how arbitrary one's method of weighting is. Assigning equal weights regardless of methodological quality is maximally arbitrary. The use of QATs to differentially weight primary-level studies is an attempt to minimize such arbitrariness. Second, as argued in section "Quality assessment tools" above, one must account for fine-grained methodological features in order to guarantee that one avoids potential defeating properties of evidence, and QATs can help with this. Third—but closely related to the second point—there is some empirical evidence which suggests that studies of lower quality have a tendency to over-estimate the efficacy of medical interventions (see footnote 8), and thus the use of QATs helps to accurately estimate the efficacy of medical interventions.[14] In short, despite their low inter-rater and inter-tool reliability, QATs are an important component of medical research, and should be employed when performing systematic reviews.

Underdetermination of Evidential Significance

The primary use of QATs is to estimate the quality of evidence from particular medical studies, and the primary use of such evidence is to estimate the strength (if any) of causal relations in relevant domains.[15] The best available QATs appropriate to a given domain differ substantially in the weight assigned to various method-ological properties (section "Quality assessment tools"), and thus generate widely discordant estimates of evidential quality when applied to the same evidence (section "Inter-tool reliability"). The differences between the best available QATs are fundamentally arbitrary. Although I assume that there must be a unique value (if at all) to the strength of purported causal relations in the domains in which QATs are employed, the low inter-tool reliability of QATs—together with the fundamentally arbitrary differences of their content—suggests that, in such domains and for such relations, there is no uniquely correct estimate of the quality of evidence.

Disagreement regarding the strength of evidence in a particular scientific domain has been frequently documented with historical case studies. One virtue of examin-ing the disagreement generated by the use of QATs is that such disagreements occur in highly controlled settings, are quantifiable using measures such as the κ statistic discussed above, and are about subjects of great importance. Such disagreements do not necessarily represent shortcoming on the part of the disagreeing scientists, and nor do such disagreements necessarily suggest a crude relativism. Two scientists who disagree about the strength of a particular piece of evidence can both be ratio-nal because their differing assessments of the strength of the same evidence can be due to their different weightings of fine-grained features of the methods which gen-erated the evidence. This explains (at least in part) the low inter-rater and inter-tool reliability of QATs.

Concluding that there is no uniquely correct determination of the epistemic significance of some piece of evidence by appealing to the poor inter-rater and inter-tool reliability of QATs is not merely an argument from disagreement. If it were, then the standard rejoinder would simply note that the mere fact of disagree-ment about a particular subject does not imply that there is no correct or uniquely best view on this subject. So although different QATs disagree about the strength of evidence from a particular trial, this does not imply that there is no true or best view

---

[14] This latter consideration is somewhat controversial, both because it has been contradicted by other empirical studies, and because it assumes that the correct estimate of the efficacy of medical interventions is given by what are purported to be higher quality studies.

[15] The relata in such purported causal relations are, of course, the medical intervention under investigation and the change in value of one or more parameters of a group of subjects.

regarding whether or not the evidence from this particular trial is strong, since the best QATs might agree with each other about the evidence from this trial, and even more ambitiously, agreement or disagreement among QATs would be irrelevant if we just took into account the quality assessment of this particular trial by the uniquely best QAT. The burden with this rejoinder is to identify the single best QAT or at least the set of best QATs (and then hope that multiple users of the single best QAT will have high inter-rater reliability or that the set of best QATs will have high inter-tool reliability). As noted in section "Inter-tool reliability", medical scientists involved in the development and assessment of QATs hold that there are simply a plurality of decent QATs that differ from one another in arbitrary respects. More fundamentally, we lack a theory of scientific inference that would allow us to referee between the most sophisticated QATs. Recall the different weightings of the particular methodological features assessed in QATs noted in Table 1. Another way to state the burden of the 'mere argument by disagreement' rejoinder is that to identify the best QATs or single best QAT, one would have to possess a principled method of determining the optimal weights for the methodological features included on a QAT. That we do not presently have such a principled method is an understatement.

Consider this compelling illustration of the arbitrariness involved in the assignment of weights to methodological features in QATs. Cho and Bero (1994) employed three different algorithms for weighting the methodological features of their QAT (discussed in section "Quality assessment tools"). Then they tested the three weighting algorithms for their effect on quality scores of medical trials, and their effect on the inter-rater reliability of such scores. They selected for further use—*with no principled basis*—the weighting algorithm that had the highest inter-rater reliability. Cho and Bero explicitly admitted that nothing beyond the higher inter-rater reliability warranted the choice of this weighting algorithm, and they rightfully claimed that such arbitrariness was justified because "there is little empiric [sic] evidence on the relative importance of the individual quality criteria to the control of systematic bias."[16] Medical scientists have no principled foundation for developing a uniquely good QAT, and so resort to a relatively arbitrary basis for developing QATs.

One might press the above response by noting that while it is true that we *presently* lack an inductive theory that could provide warrant for a unique system for weighting the various methodological features, it is overly pessimistic to think that we will *never* have a principled basis for identifying a uniquely best QAT. It is plausible, this objection goes, to think that someday we will have a uniquely best QAT, or perhaps uniquely best QATs given particular kinds of epistemic scenarios, and we could thereby achieve agreement regarding the strength of evidence from medical studies. To this one would have to forgive those medical scientists, dissatisfied with this

---

[16] There is a tendency among medical scientists to suppose that the relative importance of various methodological features is merely an empirical matter. One need not entirely sympathize with such methodological naturalism to agree with the point expressed by Cho and Bero here: we lack reasons to prefer one weighting of methodological features over another, regardless of whether one thinks of these reasons as empirical or principled.

response, who are concerned with assessing evidence today. But there is another, deeper reason why such a response is not compelling.

It is not a mere argument from present disagreement to claim that the poor inter-tool reliability of QATs implies that the strength of evidence from particular medical studies is underdetermined. That is because, as the example of the Cho and Bero QAT suggests, the disagreements between QATs are due to arbitrary differences in how the particular methodological features are weighed by QATs. There are, to be sure, better and worse QATs. But that is about as good as one can do when it comes to differentiating QATs. Of those QATs that account for the majority of relevant methodological features, some weight those features in a slightly different manner than others, and there is no principled grounds for preferring one weighting over another. We do not possess a theory of scientific inference that could help determine the weights of the methodological features in QATs. If one really wanted to, one could sustain the objection by claiming that it is possible that in the future we will develop a normative theory of inference which would allow us to identify a uniquely best QAT. There is a point at which one can no longer argue against philosophical optimism. The underdetermination of evidential significance is a hard problem; like other hard philosophical problems, it does not preclude optimism.

One could put aside the aim of finding a principled basis for selecting among the available QATs and instead select QATs based on their historical performance. Call this a 'naturalist' selection of QATs. Since QATs are employed to estimate the quality of evidence from medical studies, and such evidence is used to estimate the strength of causal relations, the naturalist approach could involve selecting QATs based on a parameter determined by the 'fit' between (i) the strength of presently known causal relations and (ii) the quality of the evidence for such causal relations available at a particular past time, as determined in retrospect by the currently available QATs. The best QAT (for some given domain) would be simply the QAT with the best average fit between (i) and (ii). Such an assessment of QATs would be of some value. It would be limited, though, given a fundamental epistemic circularity. In the domains in which QATs are commonly employed, the best epistemic access to the strength of causal relations is the total evidence from all the available medical studies, summarized by a careful systematic review (which, in this domain, usually takes the form of a meta-analysis), appropriately weighted to take into account relevant methodological features of those studies. But of course, those very weightings are generated by QATs. The naturalist approach to assessing QATs, then, itself requires the employment of QATs.

The underdetermination of evidential significance is *not* the same problem that is associated with Duhem and Quine. The standard underdetermination problem—underdetermination of theory by evidence—holds that there are multiple theories compatible with a given body of evidence. The underdetermination of evidential significance is the prior problem of settling on the strength of a given piece of evidence in the first place. Indeed, perhaps an appropriate name for the present problem is just the inverse of the Duhem-Quine locution: the underdetermination of evidence by theory. Our best theories of inference underdetermine the strength of evidence, as measured by tools such as QATs.

Conclusion

An examination of QATs suggests that coarse-grained features of evidence in medicine, like freedom from systematic error, are themselves amalgams of a complex set of considerations; that is why QATs take into account a plurality of methodological features such as randomization and blinding. The various aspects of a specific empirical situation which can influence an assessment of a coarse-grained evidential feature are numerous, often difficult to identify and articulate, and if they can be identified and articulated (as one attempts to do with QATs), they can be evaluated by different scientists to varying degrees and by different quality assessment tools to various degrees. In short, there are a variety of features of evidence that must be considered when assessing evidence, and there are numerous and potentially contradictory ways to do so. Our best theories of scientific inference provide little guidance on how to weigh the relevant methodological features included in tools like QATs.

The most frequently used tools for assessing the quality of medical studies are not QATs, but rather evidence hierarchies. An evidence hierarchy is a rank-ordering of kinds of methods according to the potential for bias in that kind of method. The potential for bias is usually based on one or very few parameters of study designs, most prominently randomization. QATs and evidence hierarchies are not mutually exclusive, since an evidence hierarchy can be employed to generate a rank-ordering of types of methods, and then QATs can be employed to evaluate the quality of tokens of those methods. However, elsewhere (Stegenga forthcoming) I argue that judicious use of QATs should replace evidence hierarchies altogether. The best defense of evidence hierarchies that I know of is given by Howick (2011), who promotes a sophisticated version of hierarchies in which the rank-ordering of a particular study can increase or decrease depending on parameters distinct from the parameter first used to generate the ranking. Howick's suggestion, and any evidence hierarchy consistent with his suggestion (such as that of GRADE), ultimately amounts to an outright abandonment of evidence hierarchies. Howick gives conditions for when mechanistic evidence and evidence from non-randomized studies should be considered, and also suggests that sometimes evidence from RCTs should be doubted. If one takes into account methodological nuances of medical research, in the ways that Howick suggests or otherwise, then the metaphor of a hierarchy of evidence and its utility in assessing quality of evidence seem less compelling than more quantitative tools like QATs.

For instance, the GRADE evidence hierarchy employs more than one property to rank methods. GRADE starts with a quality assignment based on one property and takes other properties into account by subsequent modifications of the quality assignment (shifting the assignment up or down). Formally, the use of n properties to rank methods is equivalent to a scoring system based on n properties which discards any information that exceeds what is required to generate a ranking. QATs generate scores that are measured on scales more informative than ordinal scales (such as interval, ratio, or absolute scales). From any measure on one of these supra-ordinal scales, a ranking can be inferred on an ordinal scale, but not vice versa (from a ranking on an ordinal scale it is impossible to infer measures on supra-ordinal scales). Thus hierarchies (including the more sophisticated ones such as GRADE) provide evaluations of evidence which are *necessarily less informative* than evaluations provided by QATs.

Moreover, because these sophisticated hierarchies begin with a quality assignment based on one methodological property and then shift the quality assignment by taking other properties into account, the weights that can be assigned to various methodological properties are highly constrained. Specifically, if l is the number of levels that an initial quality assignment can be shifted, and p is the number of properties assessed in the overall evaluation, then the weight assigned to any property in such a hierarchy is limited to l/p. With QATs, on the other hand, the weight assigned to any property is completely open, and can be determined based on rational arguments regarding the respective importance of the various properties, without arbitrary constraints imposed by the structure of the scoring system. In short, despite the widespread use of evidence hierarchies and the defense of such use by Howick (2011), and despite the problems that I raise for QATs above, QATs are superior to evidence hierarchies for assessing the great diversity of evidence in contemporary medical research.

A group of medical scientists prominent in the literature on QATs notes that "the quality of controlled trials is of obvious relevance to systematic reviews" but that "the methodology for both the assessment of quality and its incorporation into systematic reviews are a matter of ongoing debate" (Jüni et al. 2001).[17] I have argued that the use of QATs are important to minimize arbitrariness when assessing medical evidence and to accurately estimate probabilities associated with measures of confirmation. However, available QATs vary in their constitutions, and when medical evidence is assessed using QATs their inter-rater reliability and inter-tool reliability is low. This, in turn, is a compelling illustration of a more general problem: the underdetermination of evidential significance. Disagreements about the strength of evidence are, of course, ubiquitous in science. Such disagreement is especially striking, however, when it results from the employment of carefully codified tools designed to quantitatively assess the strength of evidence. QATs are currently the *best* instruments available to medical scientists to assess the strength of evidence, yet when applied to what is purported to be the *best* quality evidence in medicine (namely, evidence from RCTs), different users of the same QAT, and different QATs applied to the same evidence, lead to widely discordant assessments of the strength of evidence.

Acknowledgements This paper has benefited from discussion with Nancy Cartwright, Eran Tal, Jonah Schupbach, and audiences at the University of Utah, University of Toronto, and the Canadian Society for the History and Philosophy of Science. I owe the title to Frédéric Bouchard. Medical scientists Ken Bond and David Moher provided detailed written commentary. All remaining errors remain mine alone. I am grateful for financial support from the Social Sciences and Humanities Research Council of Canada.

---

[17] See also Olivo et al. (2007) for an empirical critique of QATs.

References

Balk EM, Bonis PA, Moskowitz H, Schmid CH, Ioannidis JP, Wang C, Lau J (2002) Correlation of quality measures with estimates of treatment effect in meta-analyses of randomized controlled trials. JAMA 287(22):2973–2982

Bechtel W (2002) Aligning multiple research techniques in cognitive neuroscience: why is it important? Philos Sci 69:S48–S58

Bluhm R (2005) From hierarchy to network: a richer view of evidence for evidence-based medicine. Perspect Biol Med 48(4):535–547

Borgerson K (2008) Valuing and evaluating evidence in medicine. PhD dissertation, University of Toronto

Carnap R (1947) On the application of inductive logic. Philos Phenomenol Res 8:133–148

Cartwright N (2007) Are RCTs the gold standard? Biosocieties 2:11–20

Cartwright N (2010) The long road from 'it works somewhere' to 'it will work for us'. Philosophy of Science Association, Presidential Address

Chalmers TC, Smith H, Blackburn B et al (1981) A method for assessing the quality of a randomized control trial. Control Clin Trials 2:31–49

Cho MK, Bero LA (1994) Instruments for assessing the quality of drug studies published in the medical literature. JAMA 272:101–104

Clark HD, Wells GA, Huët C, McAlister FA, Salmi LR, Fergusson D, Laupacis A (1999) Assessing the quality of randomized trials: reliability of the Jadad scale. Control Clin Trials 20:448–452

Cohen J (1960) A coefficient of agreement for nominal scales. Educ Psychol Meas 20(1):37–46

Egger M, Smith GD, Phillips AN (1997) Meta-analysis: principles and procedures. Br Med J 315:1533–1537

Good IJ (1967) On the principle of total evidence. Br J Philos Sci 17(4):319–321

Hacking I (1981) Do we see through a microscope? Pac Philos Quart 63:305–322

Hartling L, Ospina M, Liang Y, Dryden D, Hooten N, Seida J, Klassen T (2009) Risk of bias versus quality assessment of randomised controlled trials: cross sectional study. Br Med J 339:b4012

Hartling L, Bond K, Vandermeer B, Seida J, Dryen DM, Rowe BH (2011) Applying the risk of bias tool in a systematic review of combination long-acting beta-agonists and inhaled corticosteroids for persistent asthma. PLoS One 6(2):1–6, e17242

Hempel S, Suttorp MJ, Miles JNV, Wang Z, Maglione M, Morton S, Johnsen B, Valentine D, Shekelle PG (2011) Empirical evidence of associations between trial quality and effect sizes. Methods research report, AHRQ publication no. 11-EHC045-EF. Available at: http://effective-healthcare.ahrq.gov

Herbison P, Hay-Smith J, Gillespie WJ (2006) Adjustment of meta-analyses on the basis of quality scores should be abandoned. J Clin Epidemiol 59:1249–1256

Howick J (2011) The philosophy of evidence-based medicine. Wiley-Blackwell, Chichester/Hoboken

Jadad AR, Moore RA, Carroll D et al (1996) Assessing the quality of reports of randomized clinical trials: is blinding necessary? Control Clin Trials 17:1–12

Jüni P, Witschi A, Bloch R, Egger M (1999) The hazards of scoring the quality of clinical trials for meta-analysis. J Am Med Assoc 282(11):1054–1060

Jüni P, Altman DG, Egger M (2001) Assessing the quality of randomised controlled trials. In: Egger M, Smith GD, Altman DG (eds) Systematic reviews in health care: meta-analysis in context. BMJ Publishing Group, London

La Caze A (2011) The role of basic science in evidence-based medicine. Biol Philos 26(1):81–98

Linde K, Clausius N, Ramirez G et al (1997) Are the clinical effects of homoeopathy placebo effects? Lancet 350:834–843

Mayo D (1996) Error and the growth of experimental knowledge. University of Chicago Press, Chicago

Mayo D, Spanos A (2011) The error statistical philosophy. In: Mayo D, Spanos A (eds) Error and inference: recent exchanges on experimental reasoning, reliability, and the objectivity and rationality of science. Cambridge University Press, New York

Moher D, Jadad AR, Nichol G, Penman M, Tugwell P, Walsh S (1995) Assessing the quality of randomized controlled trials: an annotated bibliography of scales and checklists. Control Clin Trials 16:62–73

Moher D, Jadad AR, Tugwell P (1996) Assessing the quality of randomized controlled trials. Current issues and future directions. Int J Technol Assess Health Care 12(2):195–208

Moher D, Pham B, Jones A, Cook DJ, Jadad AR, Moher M, Tugwell P, Klassen TP (1998) Does quality of reports of randomised trials affect estimates of intervention efficacy reported in meta-analyses? Lancet 352(9128):609–613

Olivo SA, Macedo LG, Gadotti IC, Fuentes J, Stanton T, Magee DJ (2007) Scales to assess the quality of randomized controlled trials: a systematic review. Phys Ther 88(2):156–175

Reisch JS, Tyson JE, Mize SG (1989) Aid to the evaluation of therapeutic studies. Pediatrics 84:815–827

Spitzer WO, Lawrence V, Dales R et al (1990) Links between passive smoking and disease: a best-evidence synthesis. A report of the working group on passive smoking. Clin Invest Med 13:17–42

Stegenga J (2011) Is meta-analysis the platinum standard of evidence? Stud Hist Philos Biol Biomed Sci 42:497–507

Stegenga J (forthcoming) Down with the hierarchies

Thagard P (1998) Ulcers and bacteria I: discovery and acceptance. Stud Hist Philos Biol Biomed Sci 29:107–136

Upshur R (2005) Looking for rules in a world of exceptions: reflections on evidence-based practice. Perspect Biol Med 48(4):477–489

Weber M (2009) The crux of crucial experiments: Duhem's problems and inference to the best explanation. Br J Philos Sci 60:19–49

West S, King V, Carey TS, Lohr KN, McKoy N, Sutton SF, Lux L (2002) Systems to rate the strength of scientific evidence. Evidence report/technology assessment number 47, AHRQ publication no. 02-E016

Worrall J (2002) *What* evidence in evidence-based medicine? Philos Sci 69:S316–S330

Worrall J (2007) Why there's no cause to randomize. Br J Philos Sci 58:451–488